Linear Integrated Circuits

Practice and Applications

Linear Integrated Circuits

Practice and Applications

Sol D. Prensky

Associate Professor Emeritus
Fairleigh Dickinson University

Arthur H. Seidman

Professor of Electrical Engineering
Pratt Institute

Reston Publishing Company, Inc., Reston, Virginia
A Prentice-Hall Company

Library of Congress Cataloging in Publication Data

Prensky, Sol D.
    Linear integrated circuits.

    Edition of the 1974 by S. D. Prensky published
under title: Manual of linear integrated circuits.
    Includes index.
    1. Linear integrated circuits. I. Seidman,
Arthur H., joint author. II. Title.
TK7874.P73 1981   621.381'73   80-26652
ISBN 0-8359-4084-5

© 1981 by
Reston Publishing Company, Inc.
A Prentice-Hall Company
Reston, Virginia

10 9 8 7 6 5 4 3 2 1

Printed in the United States of America

In memory of Sol D. Prensky
teacher, writer—a warm and generous human being

Contents

Preface

In considering the sustained progress of the field of integrated circuits, we find important major advances that have been made over a relatively short period. Most outstanding has been the dramatic development of the microprocessor (μP) in the digital IC field, accompanied by supplemental progress in the linear IC group. This significant progress has been sparked by important advances in large-scale-integration (LSI) technology in achieving considerably greater density of active devices on the IC chip, including both the bipolar and the MOS field-effect types of transistors, and also the new combination of both devices (called a BiMOS or BiFET chip). Combined with these improved fabrication techniques, we find clever circuit designs within the denser IC chip that offers a choice of various monolithic microprocessor units, which merit the proud description of "computer-on-a-chip."

This active development of microprocessors and their attendant memory circuits in the digital division has fostered a corresponding upsurge of advanced devices in the linear IC field. Exploiting the broad flexibility of the micro-processor, we find innovative designs that include linear ICs diverse areas, ranging from the automotive field for engine controls to the electronic-instruments field for various automatic testing functions. In these new functions, many advanced forms of linear ICs are employed to handle the analog signals that feed the digital processing.

Other linear IC developments, such as in the consumer-communication areas, all emphasize the continuing trend for replacing discrete circuits (with their multiplicity of individual transistors and their numerous interconnections) with the more compact and reliable ICs. In the broad view, it is fair to say that the process of supplanting discrete transistor circuits by the emergent ICs is now as much a revolutionary development as was the previous upheaval that supplanted vacuum tubes with transistors.

For the linear IC field, advanced developments include newer versions of OP AMPS (including the previously mentioned BiFET types), increased utilization of the phase-locked-loops (PLL) units in the communications area, and a proliferation of devices interfacing with the newer digital developments. To cite some additional examples, we find more versatile A/D and D/A converters, more multipliers for signal conditioning and a greater variety of peripheral drivers for various interface and display purposes.

This work is an updating and expansion of Prensky's well-received *Manual of Linear Integrated Circuits: Operational Amplifiers and Analog ICs* published by Reston Publishing Company in 1974. Its aim is to provide the technician and engineer with practical information on the selection of linear ICs for use in a large variety of applications which include dc and audio amplifiers, waveform generation, D/A and A/D conversion, active filters, voltage regulators, and communication systems.

In keeping with the original theme of representing well-accepted linear IC models from all the major manufacturers, considerable care has been taken in presenting the extensive cross-reference list in Appendix III. This listing of around 400 frequently used model numbers is an important practical feature; it allows the user, often confronted with a clutter of type numbers, to rapidly identify a particular manufacturer's designation, and also to characterize it in relation to its general class (often as a second source of a better-known type).

This same feature of including types from all the major manufacturers has been carried out in the presentation of the selection guide for OP AMPS (Appendix II). Further, as a source for obtaining greater details on particular models that are available from the comprehensive data sheets supplied by the manufacturers, a current list of their addresses is given in Appendix IV.

The maturing growth of the semiconductor industry strengthens the trend toward "standardized" types, both for OP AMPS and for the other forms of linear ICs; to this desirable end, the manual serves as a practical tool to highlight such specific types, by illustrating selected applications of these versatile devices.

The authors are grateful to the many manufacturers who supplied us with material. Any errors, however, that may appear in the text are our responsibility.

SOL D. PRENSKY
ARTHUR H. SEIDMAN

Chapter 1

Introduction to Linear Integrated Circuits

1-1 GENERAL SIGNIFICANCE OF LINEAR INTEGRATED CIRCUITS

The category of *linear integrated circuits* (LICs) can be distinguished, in a broad sense, from the category of digital integrated circuits (DICs); thus the numerous and diverse types of integrated circuits can be classified into these two large groups. Both forms of ICs (also known as *microcircuits*) have proved immensely popular, and they are rapidly supplanting circuits using discrete transistors. This stems from the greater convenience and increased reliability offered by these small devices.

The linear IC group encompasses circuits of the *analog type*, where the input, in general, is in the form of a smoothly varying signal and the output usually is an amplified version of the input signal. This is in contrast to the zero-one switching action of the digital IC. The *operational* amplifier (OP AMP) is an outstanding and popular example of the analog group, which, in addition, includes other nondigital types (even though the operation may be nonlinear). Thus the LIC group includes regulators and other forms of signal conditioners that are designated "linear" or analog, as opposed to the digital types. The DIC, on the other hand, depends essentially on its switching function, and this group encompasses a large number of devices, such as gates, flip-flops, memories, and microprocessors.

As might be expected, with the wide diversity of ICs there are some devices that possess characteristics of both groups, as in the case of *interface circuits*. Examples of interface circuits include the analog-to-digital (A/D) converter and its counterpart, the digital-to-analog (D/A) converter. In addition, important use is made of these "in-between" types in providing a great variety of control circuits using the highly versatile microprocessor, where analog sensing elements are conditioned to work with the digital processing of microcomputers. These special types are generally classified as belonging to the linear group. Nevertheless, there need be no confusion about these classifications, as long as we do not interpret the designation of linear in too literal a fashion.

Thus the linear group may include devices that exhibit nonlinear transfer characteristics but that still are of an analog nature (such as the logarithmic amplifier). Keeping in mind the broad interpretation of these classifications, this text concentrates on the entire linear group as opposed to the other major (and very large) group of digital ICs.

1-2 DEVELOPMENT OF LINEAR INTEGRATED CIRCUITS

The linear group of ICs attained its importance and prominence when it eventually became possible to provide, at a reasonably low cost, amplifiers of the *operational-amplifier type*. The development can be traced from the time (approximately 1964) when pairs of transistors, forming *integrated differential amplifiers*, were first fabricated on a single silicon chip. This step represented a significant advance over the use of discrete transistors in the basic circuit; it demonstrated the ability of the IC to greatly reduce the troublesome temperature dependence of discrete transistors, even when the separate transistors were laboriously matched (as discussed later). Since both transistors of the differential pair could be fabricated simultaneously on the same IC chip, it was possible, by this step alone, to make an improvement by whole orders of magnitude in alleviating the temperature-drift problem—from millivolts for the discrete transistor to just microvolts per degree Celsius for the integrated pair.

About a year later, another enormous step forward was taken with the introduction of the *integrated general-purpose operational amplifier*. This IC was able to take good advantage of the integrated differential-amplifier stage for the first stage of a multistage amplifier and was capable of open-loop gains well beyond 10,000—all in a very compact package. By the use of an external feedback resistor, the IC operational amplifier (of the 709[1] type, for example) became a low-cost and highly versatile amplifier that incorporated 9 transistors and 12 associated resistors in a conveniently small package. This relatively simple operational amplifier could easily perform many of the functions of discrete amplifiers and could do so at a lower cost and in a much more reliable

---

[1] A popular IC type.

and versatile form. The IC operational amplifier offered designers a highly flexible tool that approached a *basic building block* around which a great number of desired circuits could be devised quite simply.

1-3 USING A SIMPLE LINEAR INTEGRATED CIRCUIT

The ease of use and the great utility of LICs can be illustrated by selecting a simple example in the form of a general-purpose operational amplifier (commonly called an OP AMP). Contrasted with the use of discrete components (such as transistors, resistors, and small capacitors) in a multistage amplifier, the use of an OP AMP can often make the design of a desired amplification function (either direct or alternating current) a straightforward operation; in many cases it would simply mean choosing values for two external resistors (R_f and R_i) to use with a general-purpose OP AMP, such as the very popular 741 type[2] illustrated in Fig. 1.1.

To gain a clear appreciation of the superior convenience and flexibility offered by the IC OP AMP compared to a traditional amplifier made up of discrete components, it is instructive to trace the steps needed in the case of both amplifiers to accomplish a typical amplification function.

Let us assume a project requiring a direct-coupled amplifier with a stable gain of, say, 500, where the input comes from a light sensor that provides a slowly varying dc signal ranging from 1 to 10 mV. (This would call for an output from the amplifier of $\frac{1}{2}$ to 5 V, as displayed on a dc voltmeter.) In the next two paragraphs we can follow the necessary procedures to accomplish the same purpose in each case.

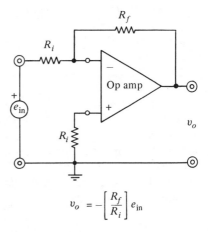

$$v_o = -\left[\frac{R_f}{R_i}\right]e_{\text{in}}$$

Figure 1.1 Functional symbol diagram of an operational amplifier (OP AMP); the simple gain formula applies to a typical operational amplifier used as an inverting amplifier.

[2] Another very popular IC type with internal compensation.

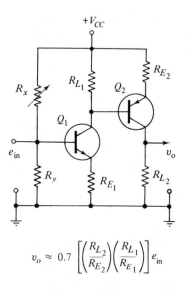

$$v_o \approx 0.7 \left[\left(\frac{R_{L_2}}{R_{E_2}} \right) \left(\frac{R_{L_1}}{R_{E_1}} \right) \right] e_{\text{in}}$$

Figure 1.2 Discrete form of direct-coupled amplifier: it requires calculation of proper values of at least six resistors to accomplish desired gain; compare this to a simple ratio of two resistors for the OP AMP in Fig. 1.1.

In the case of the *discrete amplifier*, we might choose the circuit of Fig. 1.2, calling for an *NPN* transistor (Q_1) followed by a *PNP* type (Q_2), arranged in the "compound" type of circuit, where the bias voltages can be handled more conveniently in this type of complementary transistor connection. Making use of simplified design relations, we would first establish the Q-point for the output stage to ensure that we stay within the linear operating range. With a supply voltage of around 20 V, this would call for a collector voltage of about half the supply, or approximately 10 V, for V_{CE2} at the Q-point, ensuring no distortion. Then, as suggested by Lenk,[3] we would choose proper values for the resistors; for example, we might choose R_{L_2} as 1 kΩ, requiring collector current of Q_2 to be about 10 mA. Proceeding from this, values of R_{E_2} and then of R_{L_1} and R_{E_1} are selected for the proper gain relationship of R_L/R_E for each stage; then the selection of approximate values of R_x and R_y is made for obtaining proper bias currents. As an additional simplification, R_x is shown to be variable, so that it may be easily set to obtain the correct meter reading for collector current of Q_2; finally, a final adjustment of R_x is again made, if required, so that the full output range is obtained without distortion. It will be noted that even when a bit of cut-and-try process is used here to reduce unnecessarily precise calculations, the simplified procedure still involves working with six resistor values to accomplish the proper operation of the amplifier in producing a direct-coupled gain of about 500, together with a satisfactory output swing between $\frac{1}{2}$ and 5 V, as desired.

[3] J. D. Lenk, *Handbook of Simplified Solid-State Circuit Design*, Prentice-Hall, Inc., Englewood Cliffs, N.J., 1971.

By contrast, when using the general-purpose IC OP AMP (shown in Fig. 1.1), the procedure involves only the selection of two resistors to obtain the proper ratio of $R_f/R_i = 500$. Thus R_f might be 500 kΩ, with R_i equal to 1 kΩ (the third resistor is generally made roughly equal to R_i). With these three resistors and the dual power supply (such as ±15 V) connected, the amplifier circuit is ready for the signal, as the proper bias and stability conditions have already been taken care of by the internal design of the IC. Additionally, since the IC is capable of at least a ±10-V output swing, the required output indication here (up to 5 V) is easily handled, without concern for any distortion arising from overdriving. Further advantages lie in the ease of changing the amount of gain and also, when required, in easily obtaining higher input impedances, as will be shown later.

If we now add to the preceding evidence the pertinent fact that the IC, incorporating more than a dozen transistors, practically always costs less than an equivalent discrete circuit, we can clearly see the great potential of the linear IC in *reducing both complexity and expense* in a multitude of applications. A *note of caution* is in order here for those who might jump to the conclusion that the welcome simplicity in the use of the LIC now makes it unnecessary to study and understand the design considerations of discrete transistor circuits. It should be kept in mind that to make effective use of the linear IC in its many forms it is still necessary to thoroughly understand the basic transistor principles— knowledge needed to determine the gain, stability, frequency-response, and power-capability characteristics of any transistor circuit, be it in the discrete or in the integrated form.

Although our example is admittedly a limited case of simple amplifier function (since it has been selected here particularly for its simplicity), it is still a valid one for a great many practical applications. When more specific applications are needed, there will naturally be other amplifier elements to be considered regarding bandwidth, power output, and the like; each of these desired amplifier elements will be discussed later in the light of the great variety of LICs available, including not only various types of OP AMPS, but many other linear IC types as well.

1-4 TYPES OF LINEAR-INTEGRATED-CIRCUIT FUNCTIONS

Linear ICs are designed to perform an extremely wide variety of analog functions. If we start with a very rough division of analog functions into the two main areas of *amplification* and *oscillation* (tentatively leaving the *switching* function to the digital ICs), we find that the analog functions call for many subdivisions of amplifier types. Citing just a few, for example, we have *low-level* direct-coupled amplifiers for instrumentation, *wide-band* amplifiers for general

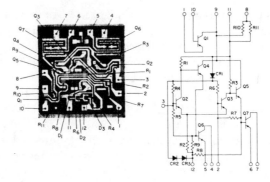

Figure 1.3 Enlarged internal view of monolithic integrated circuits showing
component location and circuit diagram (KD2115); actual size of
IC chip is less than 0.16 cm (1/16 in.) square. (*RCA "Solid-State
Hobby Circuits Manual"*)

use, and *tuned amplifiers* (with associated *power amplifiers*) for use in communication-type circuits.

In each of these cases, the circuit design can take different forms, as
follows:

1. Using *discrete components*, in the traditional method.
2. Use of a *monolithic IC*, where both active and passive components are
    fashioned on a single chip. (See Fig. 1.3.)

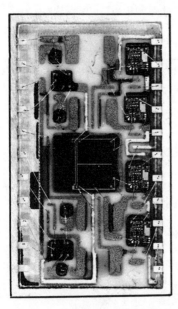

Figure 1.4 Enlarged internal view of hybrid (modular) IC. (*Hybrid
Systems Corporation*)

3. Use of a *hybrid IC*, involving a combination of chips or processes in a single package. (See Fig. 1.4.)
4. A *combination* of any of these forms (generally called a *hybrid circuit*, as distinguished from a hybrid IC).

Although the most effective form, of course, depends upon the particular circuit application, the great convenience and potential versatility of the *monolithic linear IC package* have made it a preferred choice in those cases where it is possible to develop a circuit of general utility in a practical and economical form. As a result, a host of generally useful monolithic linear ICs has been made available in the variety of types discussed in Section 1-5. Hybrid ICs and combinations with discrete components are also exemplified later in those instances where more specific applications identify them as the most effective choice.

1-5 TYPES OF LINEAR INTEGRATED CIRCUITS

A first glance at any long list of presently available linear ICs reveals that the greatest number of them fall into the group of operational amplifiers (OP AMPS). There are also at least five other sufficiently numerous groups that warrant consideration in an initial classification. These six main groups may be classified as follows:

1. Operational amplifiers.
2. Comparators.
3. Audio amplifiers.
4. Wide-band and radio-frequency circuits.
5. Voltage regulators.
6. Digital-interface circuits.

The characteristics and applications of each of these groups (including their subdivisions) will be discussed, in turn, in separate chapters. It is enough at this point, to appreciate the wide versatility of linear IC types available in the six groups, plus some others that are covered in Chapter 12. This text does not attempt to cover the equally numerous field of digital ICs widely used in computer and other pulse-circuit applications. However, the linear ICs that are employed to interface with digital circuits are covered in a separate chapter, as the sixth of the aforementioned groups.

1-6 MEDIUM-SCALE AND LARGE-SCALE INTEGRATION

The emergence of newer (and more complex) IC types is directly related to the progressively larger number of elements that can be successfully integrated in the single and multiple chips that make up the IC packages. The previously mentioned 741 type, for example, involves the integration of 22 transistors and over 12 associated resistors and is called *medium-scale integration (MSI)*. Moreover, the tendency to include larger numbers continues in the development of *large-scale integration (LSI)*, especially in the digital field. The designation of LSI has come to mean the integration of over 100 active components in the resulting IC package. Presently, most linear ICs fall into the MSI class of integration.

1-7 STANDARDIZATION TENDENCIES FOR LINEAR INTEGRATED CIRCUITS

Standardization of type numbers for semiconductor devices, in general, although highly desirable in theory, has always been a difficult goal to achieve in the highly competitive electronics industry—as evidenced by the fact that over 6,000 registered types of 2N transistors alone are presently available, to say nothing of the thousands of registered numbers of the diode–1N type. However, there is a more hopeful indication of a developing tendency in the linear IC field, in that a growing number of manufacturers are producing one another's popular types (or *second-sourcing* such ICs). This becomes apparent from the growing practice by manufacturers of retaining the type number of a particular IC type in their own type number, usually preceded by their own identifying initials. For example, the aforementioned 741 type of *internally compensated* OP AMP can be obtained from various manufacturers under such numbers as μA741 (originally Fairchild) and second-sourced as shown in part in Table 1-1.

TABLE 1-1

Similar (Second-Sourced) Type Numbers of Various Manufacturers[a]

μA741 (Fairchild)	CA3741 (RCA)
MC1741 (Motorola)	S5741 (Signetics); now μA741
LM741 (National)	SN52741 (Texas Instruments)

[a]A selected but more complete list of linear IC manufacturers (and addresses) is given in Appendix IV.

Many other examples of this second-sourcing process may be cited; for example, the dual OP AMP CM1458/1558 (Motorola) is second-sourced as CA3458, CA3558; and superbeta OP AMP LM108 (National) is second-sourced as SN55108. Other examples of types being standardized are summarized later.

It will be noted that these examples of this tendency toward standardization have included only six of the dozens of IC manufacturers. It would not be fair to conclude from this that many other firms do not produce important and innovative IC designs. A list of the more generally known *IC firms and their addresses is given in Appendix IV*, and many useful and significant *data sheets* and *application notes* can be obtained from them; such freely given information will be found to be of great help in rounding out the picture.

Within the limitations of this text, it is naturally necessary to limit our discussion to the typical examples that represent a given device irrespective of a particular manufacturer. There are, of course, many other special and desirable types. Such information must be obtained from other sources, such as those mentioned in Section 12-15; a comprehensive compilation of LICs from practically all manufacturers, with periodically updated issues, is the *D.A.T.A.* and *IC Updated Master* publications.

1-8 MAKING A PRACTICAL CHOICE OF AN INDUSTRY-STANDARD OP-AMP TYPE

A most practical way of establishing a firm base for working with the mass of presently available LIC models is given in the Appendices at the end of the book. Appendix II gives a *highly selective list for choosing a fairly standardized* OP-AMP type for a particular purpose, while a *comprehensive listing for identifying over 400 current LIC models* is given in the updated cross-reference list in Appendix III.

1-9 SEQUENCE OF PRESENTATION

When considering the various types of linear ICs, the subgroup of OP AMPS comes first to mind as the type most widely used in a multitude of versatile amplifier applications. The justly merited popularity of the OP AMP, however, should not be allowed to overshadow the great usefulness of the other types of LICs.

Accordingly, after a thorough treatment of OP AMPS in the first half of the manual, subsequent chapters discuss the other main LIC groups listed in Section 1-5. Some of these groups are easily identified, such as comparators, regulators, and special kind of amplifiers. Others are lumped under general area headings, as in the case of consumer-communication and digital-interface circuits.

1-10 LINEAR SUBSYSTEMS

Of special interest are the LIC types that form subsystems where a number of functions are combined on one chip. For radio and television use, for example, we have a single LIC acting as a subsystem to perform both FM limiting and detection functions, as discussed in Chapter 8. As another example, we find a LIC used as a subsystem in a microprocessor system for analog-to-digital (A/D) and digital-to-analog (D/A) conversions, described in Chapter 10.

As an aid in identifying particular type numbers encountered in the wide diversity of LICs, Appendix III provides the manufacturers' designations and descriptions of hundreds of industry accepted devices, which are listed in alphanumeric order by manufacturers' type-number designations.

Chapter 2

General Conditions Governing Linear-Integrated-Circuit Design

2-1 MONOLITHIC VERSUS DISCRETE COMPONENTS

Mass-produced monolithic integrated circuits can provide large cost savings when replacing conventional "discrete" electronic components such as transistors, diodes, resistors, and capacitors; and this cost advantage increases as the monolithic circuit is made more complex, since added components on an integrated circuit do not increase its cost of manufacture proportionately. Mass production of linear ICs relies upon the ability of LIC designers to use a basic chemical and photographic process originally intended for the making of silicon discrete transistors. Advances in this technique make it possible now to simulate the characteristics of the other active and passive components needed for complete circuit operation, to reproduce more than a single silicon transistor on the same chip, and to provide an interconnection pattern—all forming the marvelously compact IC device.

Because the process is optimized for best transistor characteristics, other needed components are usually a practical compromise. In some aspects, such as tolerance, breakdown voltage, and range of available values, monolithic components are inferior to inexpensive discrete components. For this reason, the earliest ICs to achieve wide usage were the less critical *digital-computer types*, as their on–off switching functions did not depend heavily upon accurately controlled components. Early attempts to build LICs simply as copies of existing discrete designs resulted in LIC types that were marginally producible with great care in controlling the monolithic process, but that could never be manufactured so easily as to allow the necessary low pricing of today's LICs.

Modern LICs are the result of new circuit design techniques, developed over several years, that are less dependent upon the uncontrollable characteristics of the IC process and capitalize on the unique advantages of microscopic, photographically reproduced elements. Among these advantages are the *inherent precise matching* of identical adjacent elements; the improved frequency response allowed by microscopic, *low-capacitance interconnections* between elements; and the availability of *multiple-transistor structures*, not limited by discrete-transistor cost considerations. Moreover, unique elements have been developed, using simple photomask-drawing techniques as well as slight variations in the standard IC process, which have no parallel in discrete components and, in some cases, which can only become useful when the inherent matching of a monolithic circuit is available. These elements include the *superbeta transistor* and *multiple structures*, such as the multiple-emitter *NPN* transistor and multiple emitter–multiple collector *PNP* transistors. Additionally, *compatible field-effect transistors* and other new structures are continually evolving from new LIC requirements.

2-2 MONOLITHIC *NPN* TRANSISTORS

The technology of monolithic LICs has evolved from the development of modern epitaxial, diffused, photochemically fabricated discrete transistors. Such devices are produced by the successive localized placement in a single-crystal silicon wafer of *P*-type and *N*-type material to form collector, base, and emitter regions, with *PN* junctions between them. A comparison of a discrete and monolithic *NPN* transistor structure (Fig. 2.1) shows that the structures are essentially the same, having an insulating layer of glass material above, through which contact holes are etched and aluminum connections are made to external circuitry. The monolithic transistor, however, is surrounded on the sides and bottom by a *P*-type region, as is its neighbor transistor. Discrete transistors are physically separated, after being produced thousands at a time, on a single wafer. To achieve electrical separation in the IC, a potential is applied to the surrounding *P*-type *isolation* region, which reverse biases the *PN* diode junction between it and each transistor's collector region, giving an effective open circuit between adjacent transistors. Similar isolation is provided between all other elements in the IC.

Thus, for discrete and monolithic *NPN* transistors having the same collector, base, and emitter shapes (as viewed from above), electrical characteristics are remarkably similar—but not the same. Whereas the discrete transistor has only three terminals, the monolithic transistor now has four: emitter, base, collector, and isolation. While the isolation is reverse biased under normal operating conditions, it still has an added parallel capacitance to the collector, which can

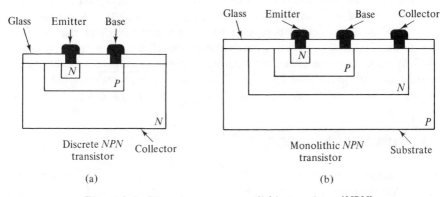

Figure 2.1 Discrete versus monolithic transistor (*NPN*).

degrade high-frequency response. Furthermore, the four-layer *NPNP* monolithic transistor structure is the same as the four-layer structure of a silicon-controlled rectifier, which is known to be a *latching device*. Normally, the fourth or isolation layer remains reverse biased, and the latch mechanism is inoperative. But in some practical applications users of LICs may inadvertently provide external bias voltages that permit *latch-up*, sometimes causing internally destructive currents to flow. (The input stage of the first-generation popular-type 709 OP AMP is notoriously subject to this failure mechanism, which is never shown on circuit schematics by its various manufacturers; it occurs when either input is raised above the positive supply voltage.)

Another difference visible in Fig. 2.1 is that collector current flows vertically downward from the emitter of the discrete transistor, through the collector, and out the bottom of the structure. Because of the monolithic transistor's isolation, collector current must flow down from the emitter, then sideways through the collector region, then back up to a top collector contact. Of course, there is more series resistance in this longer path; thus monolithic transistors almost invariably have higher saturation resistances than identical discrete counterparts and consequently are *less efficient in handling power*.

Despite their limitations, monolithic *NPN* transistors are quite competitive in performance with discrete transistors. With appropriate precise photomask techniques, they can achieve microwave frequency response or, with larger geometries, can handle amperes of current. High-voltage processes allow breakdown into the hundreds of volts, usually at the expense of current gain or saturation resistance, just as with discrete transistors. Unlike randomly selected discrete transistors, monolithic *NPN*s can be relied upon to have very *closely matched characteristics* for current gain (h_{fe}), forward emitter–base voltage (V_{BE}), and other characteristics, allowing designs that depend for their operation upon such precise matching. After all, if the masks used to photoreproduce the transistors are geometrically matched, and if the transistors on a wafer all undergo identical chemical exposures, matching is assured. Because the monolithic *NPN* is the closest structure to its discrete counterpart of the various monolithic elements, and because inherent matching allows effective LIC design, the *NPN* transistor dominates most LIC mask layouts in number. Indeed, some existing LIC designs consist entirely of transistors, with no resistors and other elements—a design technique that would be costly with discrete devices.

2-3 MONOLITHIC DIODES

Since a diode is formed from any *PN* junction, there are several regions in the monolithic structure that may be used. The key to feasibility and economy of the LIC is that all elements are formed from the same process steps as the *NPN* transistors, the only difference being the photomask geometry and the electrical function to which each region is subjected. Available diodes, seen in Fig. 2.2, include the (a) collector–isolation, (b) collector–base, (c) emitter–base, and (d) aluminum–collector diodes.

The *collector–isolation diode* offers high breakdown voltage, but is of limited use, since the isolation side is common to the LIC's most negative supply potential in most applications.

The *collector–base diode* offers breakdown voltages nearly as high as that of the collector isolation, but it has the added advantage of being isolated from other diodes or elements in the IC. Its use is unfortunately restricted by the existence of a *parasitic* unwanted *PNP* transistor structure formed by the *P*-type diode anode, *N*-type cathode, and the *P*-type isolation. Any time three such regions exist in mutual contact, with one junction forward biased and the other reverse biased, transistor action will take place. Thus, when the collector–base (CB) diode is forward biased, the isolation's reverse biasing does no good, as parasitic current will flow into the isolation. Thus the CB diode is of limited usefulness also.

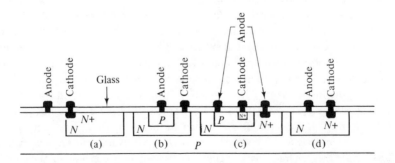

Figure 2.2 Four types of diode: (a) collector–isolation; (b) collector–base; (c) emitter–base; (d) aluminum–collector (*Schottky*).

The *emitter–base (EB) diode*, operated within an isolated N-type collector region, is also capable of parasitic current flow, by the turning on of the SCR four-layer latching mechanism. If, however, the P-base material is deliberately shorted to the surrounding N-collector region, such parasitics are prevented. Thus the most widely used diode structure in LICs is the *emitter-base, which is really a transistor, with collector and base shorted together.* Unfortunately, optimum *NPN* transistor processing requires an EB reverse breakdown between 5 and 7 V, but most LIC diodes operate in portions of the circuit where this is all that is needed. Moreover, the low reverse voltage is reliably determined by the process, allowing the EB diode to be used as a monolithic *zener diode*.

Another of the diode structures to be used is the *aluminum collector* or *Schottky diode*. Aluminum is a natural P-type doping material. When an aluminum contact is normally made in an LIC to a collector region, it is preceded by an extra heavy dose of N-type doping (indicated by $N+$), so that the interconnect contact will be *ohmic*, that is, a low-resistance contact, rather than forming a PN junction. By eliminating the heavy N contact region and maintaining very clean chemical processes, a deliberate *Schottky* or *hot-carrier diode* is formed. While having a low reverse breakdown, like the EB diode, the Schottky diode has a much lower forward drop (about 0.2 rather than 0.7 V for a silicon PN junction), and has very fast recovery time. This structure, completely within the normal LIC process, has found use in nonsaturating logic gates, fast-recovery comparators, low-power FM-IF amplifiers, and high-frequency mixers and limiters.

It should be remembered that monolithic diodes, like monolithic transistors, have extra capacitance to the isolation region, which may degrade their performance in comparison to two-terminal discrete diodes.

2-4 MONOLITHIC *PNP* TRANSISTORS

While it is easy to make a discrete *PNP* transistor (simply reverse the sequence of chemicals to make P-type and N-type regions), the monolithic *PNP* must share its process with the *NPN*. Unfortunately, if the process is best for *NPN*s, it makes rather poor *PNP*s. LIC designers have learned to live with and design around the monolithic *PNP*'s deficiencies, and, in some cases, have used it for functions that discrete *PNP*s cannot perform.

The *vertical PNP* (Fig. 2.3) is formed from the base, collector, and isolation regions, and is the same as the unwanted parasitic transistor encountered with CB diodes. Because the "doping" levels are optimized for *NPN*s, it usually has a low current gain of less than 50, and may be used only in grounded-collector (or emitter-follower) configurations, since its collector is also the common isolation region. Thus it is useful only as an emitter follower. Because of the thickness of its base region (ordinarily the *NPN* collector), its frequency response is also poor compared to the *NPN*.

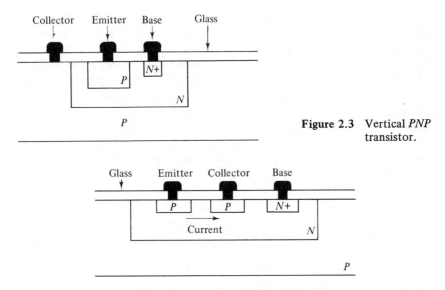

Figure 2.3 Vertical *PNP* transistor.

Figure 2.4 Lateral *PNP* transistor.

The *lateral PNP* (Fig. 2.4) has both emitter and collector formed from the same material as the *NPN*'s base, and its base from the *NPN*'s collector *N*-type region. As it is an isolated transistor, it can be used in any configuration, but because its emitter–collector spacing, dictated by photomask resolution, is wide, it has poor current gain and frequency response. LIC processing and mask layout techniques have improved upon early lateral *PNP*s to the point where they are universally used in dc and audio LIC devices, and in the dc biasing sections of high-frequency LICs.

2-5 MONOLITHIC RESISTORS

If one were to cut out a chunk of any material, such as the silicon from which LICs are made, and measure between two probes at different spots, some resistance could be determined, depending upon the resistivity of the basic material, the impurities in the material, and the size and spacing of the probe contacts. In monolithic circuits the objective is to use the available resistivity of regions to optimize the *NPN* transistor performance, without having to add or adjust processes to achieve desired resistor values. In fact, resistivities can vary considerably in the LIC process, so that it is not in the interest of mass production to expect resistors made at the same time as transistors to have accurate values.

A monolithic resistor is made by contacting two points in a region and isolating that region from everything around it, typically with a reverse-biased junction. Most popular is the base resistor; other types, such as emitter resistors,

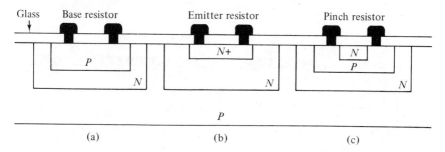

Figure 2.5   Three types of monolithic resistors: (a) base; (b) emitter; (c) pinch.

collector resistors, and *pinch* resistors, are used less frequently (see Fig. 2.5). The value of the resistor increases with increased length or with decreased width. Photochemical resolution limits minimum width so that, for a given resistivity in a region, very large resistances take up very large lengths and are uneconomical. Similarly, very low resistances require spacing of contacts too close together; thus, there are limitations to both the maximum and the minimum resistance that is available. Base resistors are generally between 100 Ω and 20 kΩ. Emitter resistors allow lower resistance, from 2 to 100 Ω, while collector resistors can reach several hundred thousand ohms. Even higher values are achieved with pinch resistors, in which a base resistor is partially pinched off by an intervening emitter region, just as a rubber water hose is pinched off.

The base resistor offers the best predicted accuracy, but even this is only ±20 or ±25 percent from a nominal value. The other types can vary ±50 or ±95 percent in some situations, so circuit biasing that is established by their value is not recommended. The higher-valued collector or pinch types are often used where a noncritical "bleeder" resistance is needed, while the very low valued emitter type is sometimes used for degeneration, parasitic-oscillation suppression, or to allow mask layout topologies where two metal strips must cross. In the latter case the emitter resistor is used as a "cross-under" and is usually not shown in the circuit schematic by the LIC manufacturer.

While circuit designs relying on absolute resistor values are not advisable in LICs, extensive use is made of the matching of two adjacent resistors. Two resistors might be off by 25 percent, but they will be within 1 or 2 percent of each other. This is especially useful in the widespread use of *differential amplifiers*, where resistor matching, not absolute value, is essential.

2-6 MONOLITHIC CAPACITORS

While junctions and resistors can be made as small as the photographic processes allow and still compete with larger discrete components, capacitor values depend primarily on their area, given available dielectric constants. Because of this, *only*

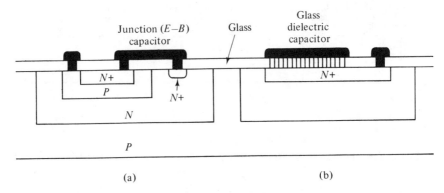

Figure 2.6 Monolithic capacitors: (a) junction type; (b) oxide type.

very small capacitors are possible in monolithic form, usually less than 200 pF. Two types are compatible with the monolithic process and are shown in Fig. 2.6. The simplest is the *junction capacitor*, in which the natural capacitance between *P* and *N* regions is used. Any junction can be used, but the EB junction gives highest capacitance per unit area, although at the 5- to 7-V EB breakdown voltage.

The capacitance of any semiconductor junction varies inversely with applied reverse voltage; thus the monolithic capacitor may be used as a voltage-variable element in oscillators and tuned circuits.

The second type, the *oxide capacitor*, uses the silicon as its lower plate; the natural insulator, silicon dioxide (ordinary sand component), which covers the IC, as the dielectric; and part of the aluminum interconnection as the upper plate. It has a high breakdown voltage, but may be hard to produce in volume unless the silicon-dioxide part of the LIC process is unusually free of localized *pinhole* defects, any one of which can short out the capacitor.

At least one "plate" of either type of capacitor is always in close contact with the isolation or some other region on the device; thus such capacitors have an unwanted, or *parasitic*, capacitance to these regions, which sometimes degrades performance when compared with discrete capacitors. Monolithic capacitors are used primarily in frequency shaping of operational amplifiers and as RF bypasses, but of which are relatively noncritical of the uncontrollably wide tolerances, typically ±50 percent.

2-7 MONOLITHIC COMPONENTS WITH NO DISCRETE COUNTERPARTS

Newer LICs contain unique structures that have resulted from the constant search to circumvent the weaknesses of existing monolithic components. Since creation of a new transistor or other structure is often no more than a mask-layout exercise using existing processes, it is easy to see that an *NPN* transistor with a single emitter region may be redrawn to contain *two or more emitters* (the basis for the popular TTL logic family); or, again, if a lateral *PNP* contains two *P*-type regions designated emitter and collector by the biasing connected to them, it is a simple matter to add one or more *P*-regions for use either as *multiple emitters or multiple collectors*. *Superbeta NPN transistors* are ordinary *NPN*s in which the emitter diffusion has been extended to create a very shallow base region, giving h_{fe} from 2,000 to 10,000, but at the price of a very low collector-emitter breakdown voltage (often less than 1 V) and very small collector currents.

The versatility of the basic monolithic LIC process has by no means been fully exploited. Components will continue to appear in mass-produced new products, which will seem to defy conventional circuit analysis, yet are only another clever variation on the available standard LIC process.

2-8 BASIC TECHNIQUES IN THE DESIGN OF LINEAR INTEGRATED CIRCUITS

Although both conventional and unique components are used in the design of the actual LIC, some basic rules can be summarized here. Older designs, in which fewer of these rules were understood, are consequently harder to manufacture, cannot give large eventual price reductions, and often are restricted in performance capability. Briefly, a well-designed LIC uses the following techniques:

1. Design relies on *matching* rather than absolute value of the components.
2. *Transistors are substituted* for resistors, diodes, and other components whenever possible. Transistors are the most controllable element available; they take less space than most elements replaced, and are inherently matched.
3. *Use differential structures.* If the biasing and other characteristics of a single-stage amplifier vary considerably because of uncontrolled components, it may be balanced against an identical amplifier that varies in the same way. (This is discussed in detail in Chapter 3, showing that an output sensitive to the difference between the two halves of the differential structure will tend to cancel out the effect of variations.)
4. If a more complex structure, such as a *compound transistor*, can give better results than a simple structure, use the more complex. It probably will cost less, all things considered.
5. Since component cost is not a consideration, the *combination of two components*, each with a weakness, can give performance superior to that possible with one premium component. For example, when a very high current gain is needed in conjunction with high breakdown voltage, as in the input of an operational amplifier, the combination of a

superbeta low-breakdown transistor with an ordinary low-beta, but high-voltage, transistor in a cascade connection gives the best of both characteristics.

6. At some point, the performance of a monolithic component, such as a large capacitor, is uneconomical compared to a discrete component. Properly designed LICs have left the monolithically impractical components to be added *external to the IC.*

2-9 MONOLITHIC VERSUS HYBRID INTEGRATED-CIRCUIT DESIGNS

Most of the statements in this book apply to monolithic LICs, that is, to circuits mass produced in a single silicon chip [mono, single; lithos (Greek), stone or crystal]. Another family of microcircuit devices, sometimes imprecisely called integrated circuits, is the *hybrid type* in which individual, microscopic components are placed on a single insulating substrate and connected by screened (thick) or evaporated (thin) conductive films. Similar films are used for resistors. While the external package may be the same size as the package in which monolithic LICs are housed, the internal size is quite different, as is the cost of construction.

A hybrid circuit can be constructed with exactly the same design techniques as an ordinary discrete-component circuit and with the same availability of precise, controlled components. It need not have any of the peculiar parasitic problems of the monolithic circuit. In general, however, it will be more expensive than a larger-sized discrete circuit because of the precise microscopic assembly labor required, and will not be as practical to repair, for the same reason. The monolithic IC, however, cannot be repaired at all. It's cheap enough to throw away if defective, and its probability of failure is much lower than that of the hybrid.

Why consider hybrids at all, then? Simply because they offer a size reduction comparable to monolithic circuits but with performance comparable to discretes. In some cases, as for *very high power* or for *high frequencies*, the hybrid can perform better than a monolithic circuit. But the performance advantage depends on the state of the monolithic art. One-watt audio power amplifiers, for example, were only feasible as hybrids in the mid 1960s. Today, 1-W and higher-power monolithic devices are readily available as LICs.

2-10 LINEAR-INTEGRATED-CIRCUIT ARRAYS

Because the *monolithic process* is governed by photoreproduction techniques and is inherently one of *duplication*, LIC manufacturers can easily build dual, triple, or higher combinations of identical or similar functions into a package. Manufacturing economics have reached the point where the cost of a single LIC die or chip is much less than the price it is sold for in packaged form, and in many instances the material and labor cost of packaging into a hermetic or plastic container exceed the cost of manufacturing even a complex die.

Simplest of these combinations is the *transistor array*, in which several transistors on one die are enclosed in a multilead container costing less than the same number of transistors individually packaged and, in certain applications, offering superior performance. More complex circuit functions, such as operational amplifiers, voltage regulators, comparators, and power amplifiers, are now also found in pairs, triples, or sometimes quadruples; each of these functional types is discussed later in separate chapters. This chapter will limit itself to the structurally simpler transistor arrays, which offer circuit designers not only cost reduction, but enable "monolithic"-type discrete circuits to be constructed, using their inherent matched characteristics.

2-11 TRANSISTOR ARRAYS

Consisting of groups of monolithic transistors, with some or all terminals available to the user, *transistor arrays* have become popular in constructing special-purpose linear circuits that may not be available in completely monolithic form, but in which monolithic design techniques are useful, at costs competitive with the use of discrete transistors. Transistor arrays can replace hand-matched dual transistors in many critical applications, such as differential-amplifier input stages, at much lower cost. While some arrays, such as the CA3045/6 and CA3018/18A,[1] make nearly all transistor terminals available externally, more efficient pin usage with a large number of array transistors usually requires internal commitment of several terminals, giving less universal arrays (such as the CA3026), which are nevertheless extremely versatile (see Figs. 2.7, 2.8, and 2.9).

Other transistor array types are aimed at specific extremes of performance. For example, types CA3081, CA3082, and CA3083[2] are high-current large-geometry transistors for display-driver applications, with correspondingly poor high frequency response. Type CA3084 is made of lateral *PNP* transistors where such devices are needed and offers excellent matching, but with relatively poor frequency response and dc current gain (see Figs. 2.10, 2.11, and 2.12). Other available arrays include zener diodes (Fig. 2.13) and thyristors (Fig. 2.14).

---

[1] High-voltage versions are CA3145E/AE, CA3118T/AT.
[2] Maximum current is 100 mA; for higher current, CA3724/5 is 1 A maximum. A higher voltage version is CA3183.

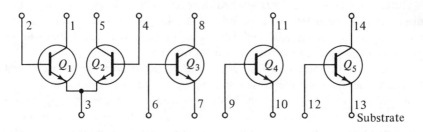

Figure 2.7   CA3046 five–transistor array. (*RCA Solid State Division*)

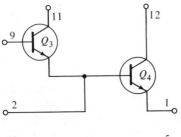

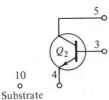

Figure 2.8   CA3018/18A four–transistor array. (*RCA Solid State Division*)

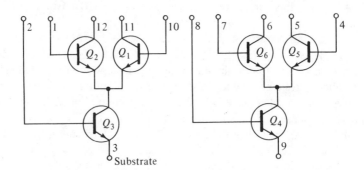

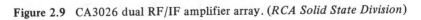

Figure 2.9   CA3026 dual RF/IF amplifier array. (*RCA Solid State Division*)

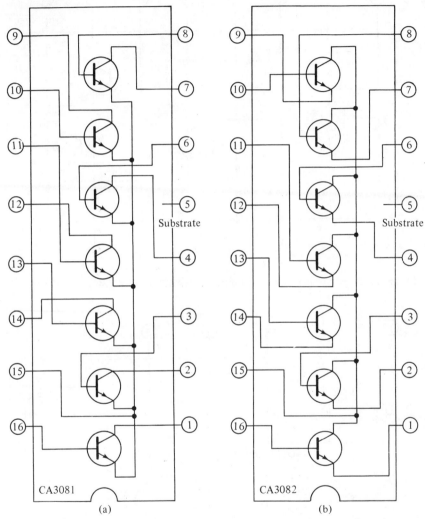

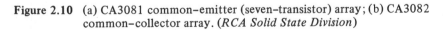

Figure 2.10 (a) CA3081 common–emitter (seven–transistor) array; (b) CA3082 common–collector array. (*RCA Solid State Division*)

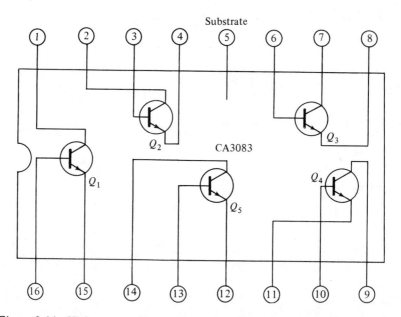

Figure 2.11 High-current (five–transistor) CA3083 array. (*RCA Solid State Division*)

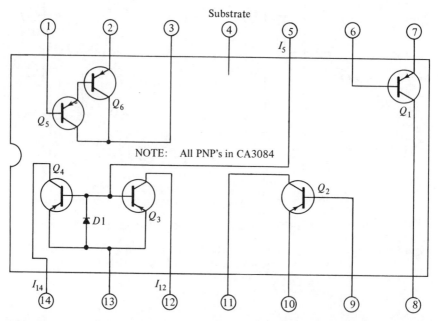

Figure 2.12 *PNP* array CA3084 containing Darlington current–mirror and two independent transistors. (*RCA Solid State Division*)

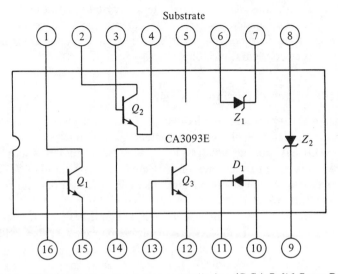

Figure 2.13 CA3093 array containing zener diodes. (*RCA Solid State Division*)

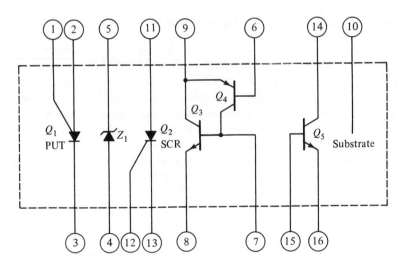

Figure 2.14 CA3097 thyristor/transistor array. (*RCA Solid State Division*)

2-12 MONOLITHIC TECHNIQUES WITH TRANSISTOR ARRAYS

By observing certain precautions, the transistor array may be used to directly replace a number of discrete transistors in a conventional linear circuit. There are configurations, however, that would simply be impractical in discrete form because of required matching, which are common in modern LICs, and are useful design tools in otherwise discrete component circuits. Three important examples of such configurations follow.

Accurately Matched Differential Pair

Many of the assumptions in the discussion of differential amplifiers (Chapter 3) rely upon ideal matching of the input transistor pair with respect to current gain (h_{fe}), forward base-emitter voltage (V_{BE} value), and input bias currents (I_B). A pair of discrete transistors randomly selected from a handful of devices, even from a single production lot from a given manufacturer, will have no better than ±20 percent current-gain match and V_{BE} offset of ±50 percent. Careful manual selection is necessary when building differential amplifiers from discrete transistors; even then, since the devices are in different packages, they will exhibit subsequent mismatches because of temperature differences and different rates of aging. Using a pair of transistors from a transistor array gives better than ±5 percent current-gain match and typical V_{BE} offset of less than ±2 mV.

Constant-Current Source or Current Mirror

Predictable constant-current sources are needed, not only in differential amplifiers, but in many other LIC functions, such as oscillators, RC timers, and the like. Since a discrete-component current source cannot rely on matched transistors, it must use precision-matched resistors to "swamp out" current-gain and V_{BE} variations seen in randomly selected discrete transistors. Such a discrete circuit is seen in Fig. 2.15. A much simpler and more effective current source is widely used in LICs, as in Fig. 2.16. For a given base–emitter structure, a fixed base–emitter voltage is required to force a fixed collector current. If transistor Q_1 is connected as in Fig. 2.16, with collector shorted to base, the Q_1 collector

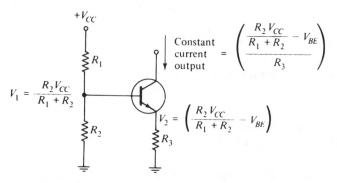

Figure 2.15 Discrete constant-current source.

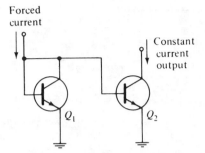

Figure 2.16 LIC current mirror.

current will increase until the sum of collector and base currents equals the forced external current, reaching an equilibrium through negative feedback. Under this condition (neglecting base current, which is much smaller than collector current), the emitter-base voltage is precisely that value required to sustain a collector current equal to the forced value. If, now, a perfect replica of Q_1, called Q_2, is connected so that the equilibrium base-emitter voltage of Q_1 is applied across the Q_2 base-emitter junction, then Q_2 must also sustain a collector current equal to the original forced value.

It may be seen that operation of the LIC constant-current source depends heavily on precise matching. This convenient structure ensures equality of currents in a transistor pair (sometimes called a *current mirror*). (The constant-current source is discussed further in Chapter 3.)

Biasing of Gain Stages with Predictable Diode (V_{BE}) Drops

A conventional discrete class A amplifier stage (Fig. 2.17) uses a number of resistors to establish its quiescent operating point. In particular, the emitter resistor (which requires a bypass capacitor if maximum gain is to be attained) establishes emitter current despite the unpredictable value of the forward base-emitter (V_{BE}) voltage drop of a randomly selected discrete transistor. An equiva-

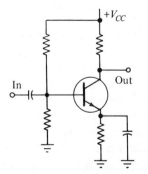

Figure 2.17 Conventional discrete amplifier.

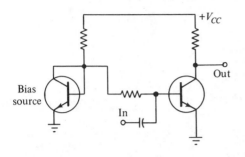

Figure 2.18   LIC amplifier using matching transistors.

lent monolithic-style amplifier (Fig. 2.18) has a grounded emitter, eliminating the emitter bypass capacitor and emitter resistor. Again, biasing such a stage by a fixed base-voltage with discrete transistors would cause unpredictable variations in collector current, whereas a transistor array used as a constant-current source provides a stable bias point. Moreover, if four or five such stages are to be similarly biased, a single *reference* voltage section, consisting of the transistor with collector–base short, can supply all stages simultaneously.

2-13 PRECAUTIONS IN APPLYING TRANSISTOR ARRAYS

In using monolithic transistor arrays it must be remembered that the individual transistors are not totally independent and are subject to the parasitic problems outlined previously in this chapter. The *following precautions* must be considered, in addition to those followed in normal discrete-transistor designs:

1. *The pin connected to the substrate of the array must always be the most negative dc voltage in the circuit*, or else one or more isolated regions will become forward biased, with respect to the substrate, under some possible circuit conditions.
2. *Saturating designs*, in which the collector–base junction of the *NPN* monolithic transistor in the array becomes momentarily forward biased, can cause the unwanted parasitic *PNPN* structure to turn on and latch. Destructive currents can be prevented under such conditions by using a series base resistance on the *NPN* transistor.
3. *Power dissipation* of a transistor array is usually limited by the total dissipation limitations of the package. While one transistor may be rated at a given maximum dissipation, it is important to calculate the total expected dissipation within the package under worst-case conditions to preclude heating problems.
4. *When several transistor-array packages* are used in a design, the transistors that are relied upon for matching must be within a *common*

package. Matching of transistors between two separate arrays is no better than that between two discrete transistors.

2-14 MULTIPLE-FUNCTION LINEAR INTEGRATED CIRCUITS

Progressive development of newer LICs has led to the introduction of devices that combine more than one function in the same compact package. Later we shall discuss LICs that include two, three, and four amplifiers housed in a single DIP package of 14 or 16 leads. In other cases there will be a number of different functions integrated in a single device, such as combinations of a voltage-controlled oscillator with an amplifier and a comparator to form an entire subsystem, all in one monolithic package.

Although in most cases the effect of companion functions may be ignored, the designer must remain aware of other cases where different LIC functions in the same package may unintentionally interact. The ways in which such interaction can take place include the following:

1. *Sharing a common power-supply connection.* Supply currents into one function must pass through unavoidable lead resistance, causing a small, but often troublesome, voltage drop in that resistance and giving rise to undesired coupling. Such situations call for various decoupling schemes.
2. *Capacitive and inductive coupling.* Because LIC packages have closely spaced leads, there may be appreciable capacitance between functions in the same package. Since capacitive reactance is lower at high frequencies, capacitive coupling is especially troublesome in multiple radio-frequency amplifiers; similarly, at the higher frequencies even the inductive reactance of the physical lead lengths must be considered.
3. *Thermal interaction.* If a LIC function is temperature sensitive, and if it shares a package with others that produce appreciable and varying amounts of internal heat, interaction will take place primarily at dc and low frequencies, since heat variation in a chip is a relatively slow process.

Despite these cautions, there is a performance advantage in complex LIC functions sharing a single package, that of matching. For example, with multiple amplifiers on a single chip, it is very likely that all operating characteristics will be closely matched and will remain matched as temperature varies, because they are in intimate thermal contact. This advantage, combined with the other integrated-circuit features of convenience and reliability, tends to favor the continual effort to produce more complex LICs that serve as subsystems and out-perform their discrete counterparts.

2-15 COMBINING TRANSISTOR FAMILIES ON A SINGLE CHIP

Advances in succeeding generations of fabrication technologies have resulted in newer devices that incorporate different transistor *families* on the same chip, such as early forms of *NPN* with *PNP* transistors. One newer example is a com-bination of *NPN* with *PNP* transistors of the high-current type in the MPQ 6502 (Motorola). This device makes it possible to provide two *complementary output-amplifier stages*, since it includes two *NPN* transistors (of the generally useful 2N2222 type) with two corresponding *PNP* transistors, suitable for complementary-amplifier circuits.

In another interesting example, bipolar transistors are combined on the chip with FET types to produce the *BiFET* or *BiMOS* forms of OP AMPS. Also, other newer arrangements facilitate *interfacing with various logic families*, such as the familiar forms of TTL (or T^2L), ECL, CMOS, and integrated-injection-logic (IIL or I^2L) forms. These and other refined forms, representing improvements over previously "standardized" general-purpose forms, are dis-cussed in later appropriate chapters.

2-16 AMPLIFIER ARRAYS

Going beyond the various transistor arrays previously discussed, there is an inter-esting *array of four amplifiers* in one package, as offered by the RCA CA3048. In this quad-amplifier array, each of the four identical amplifiers has inde-pendent inputs and outputs, all on a single monolithic chip and housed in a 16-lead DIP package.

In the schematic diagram of Fig. 2.19, two of the four amplifiers (A_1 and A_4) are shown, while the other two amplifiers (A_2 and A_3) are symmetrically similar. Referring to amplifier A_1 (at the top of the diagram), it is seen to con-sist of two stages of voltage gain. The input stage is basically a differential amplifier (transistors Q_5 and Q_6) preceded by a Darlington transistor (Q_4) on the left side. The output stage consists of a combination of three transistors (Q_{11}, Q_{12}, and Q_8) connected in an inverting configuration, with Q_8 as the actual output transistor. Fixed resistance between the output of Q_8 to the inverting input of Q_6 provides negative feedback, so that the closed-loop gain can be tailored by the value of input resistance connected to inverting terminal 3. Terminal 4 is the noninverting terminal to which the input signal is normally applied. (The fact that terminal 4 is the noninverting input can be checked by

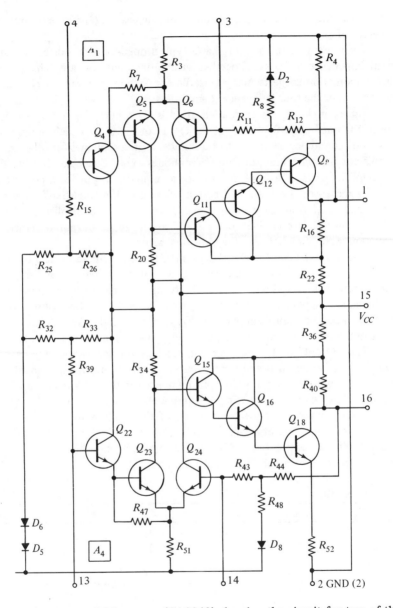

Figure 2.19 Four-amplifier array (CA3048) showing the circuit for two of the four amplifiers (A_1 and A_4); amplifiers A_2 and A_3 are symmetrically similar, and all four are on a single-chip in a 16-lead DIP package. (*RCA Solid State Division*)

noting that there is a double inversion from the base of Q_4 to the output at the collector of Q_8.)

Each amplifier in the array has a typical open-loop gain of 58 dB and an input impedance of 90 kΩ. Together with an inherent low-noise characteristic, this combination suggests many versatile applications for various compact arrangements of the four independent amplifiers.

An example of one such application is given here, illustrating a practical use of the quad-amplifier array, in the circuit of Fig. 2.20, which shows a linear mixer for combining the output of a number of microphones (or other inputs). In this circuit, two, three, or four microphones can be connected to the individual amplifier of the CA 3048 to produce a common output for a power amplifier. By means of the individual action of each of the amplifiers in the array, the sound level from each microphone, independently controlled by its own volume control, is combined or "mixed" at the single output terminal, with minimum interaction between the individual settings.

The pin-out table (or chart) shows each of the four amplifiers having the usual two input terminals, noninverting (+) and inverting (-), and a separate output terminal. Amplifiers A_1 and A_4 form one group (diagrammed in Fig. 2.20) and are supplied by the same V_{CC} 1 and ground 1, while amplifiers A_2 and A_3 form a second group, supplied by V_{CC} 2 and ground 2. Thus, if we use only two amplifiers of the array, we can avoid any power drain by the other two.

The pin-out chart shows the pin numbers for the various inputs to the noninverting (+) terminal and for the output terminals for the individual amplifiers, along with the appropriate pin numbers for the 12-V V_{CC} supply and ground. Note that the inverting (-) terminals, which are termed "bypass," serve a dual purpose. Connected to each of these minus terminals is a series combination consisting of a capacitor ($C_8 = 50$ μF in A_4) for shaping the output and a feedback resistor ($R_8 = 820$ Ω in A_4), which, together with an internal feedback resistor, completes the feedback circuit.

Pin Number Chart for the CA3048

Amplifier	Input	Output	V_{CC}	Ground	Bypass
A_1	4	1	15	2	3
A_2	8	6	12	5	7
A_3	9	11	12	5	10
A_4	13	16	15	2	14

When all four amplifiers are used, as in the mixer circuit of Fig. 2.20, the two V_{CC} terminals (15 and 12) are connected together, and similarly for the two ground terminals (2 and 5).

The amplifier gain at full volume is established for each amplifier by the value used for the individual *forward resistors* R_5, R_6, R_7, and R_8 (which are given here as 820 Ω) for an effective full gain of 10 (20 dB), with the micro-

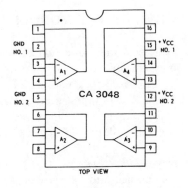

(a) pin-out diagram

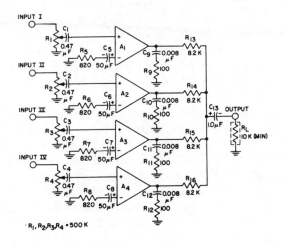

(b) mixing circuit

Figure 2.20 Quad-amplifier array (CA3048) employed for mixing the output of four microphone (or other) inputs; (a) pin terminals of 16-pin dual-in-line package (DIP); (b) circuit for single output to power amplifier. (*RCA Solid-State Application Note ICAN 4072*)

phone volume controls (R_1 through R_4) each equal to 500 kΩ. The full gain for each amplifier can be increased by lowering the value of the individual forward resistor [up to a gain of almost 500 times (or about 54 dB)] when the feedback resistor value is reduced to around 17 Ω. A curve showing this variation, along with other application details, is given in the *RCA Application Note ICAN 4072* for the CA 3048 device.

Other versions of quad-operational amplifiers from various manufacturers are discussed in Chapter 5 and later chapters, after first investigating the characteristics of single OP AMPS in the chapters that follow.

2-17 ADVANCED FORMS OF THE MONOLITHIC CHIP

Advances that have taken place in the fabrication of the analog IC chip have resulted in newer and more complex forms for the single chip. Combinations of the conventional bipolar transistor with unipolar FET types on the same chip have produced the BiFET and BiMOS types. Various forms of these types are discussed later in Section 5-17. In general, they take advantage of the *very high input impedance of the FET* with the *high speed possibilities of the bipolar devices.*

The use of the *integrated-injection-logic* (I^2L) form of fabrication has allowed greater density to be packed into a single chip. As a result of this closer packing, large-scale integration (LSI) devices have become practical for the monolithic chip, as exemplified by the many A/D and D/A converters that have been developed to accomplish the interfacing needed, for example, between analog sensors and the highly versatile microprocessor. As discussed later, a minimum of external components is required for these LSI components.

A one-chip LSI converter for a three-digit digital panel meter (DPM), for example, is available as AD2026, making an attractive simplification where greater precision is not warranted.

While the more complex monolithic forms tend to stimulate customized devices for specific applications, many of these LSI chips are being found in standardized form. One example can be seen in the automotive field, where a speed-control chip for Ford utilizes I^2L fabrication, combining an on-chip OP AMP with A/D and D/A conversion. Other examples are A/D and D/A converters in which the use of the 10-bit AD 571 *converter* makes the external component count almost negligible. For a precision instrumentation amplifier, *Signetics* offers the NE/SE-5540 in monolithic programmable form, another example of the trend of LSI permeating the analog field, as it previously did the digital field.

QUESTIONS

In the multiple-choice questions *for this chapter only*, more than one choice may be correct. Give all the correct choices.

    2-1. *Linear integrated circuits* include the following devices in addition to operational amplifiers:
          (a) Comparators
          (b) Gates
          (c) Flip-flops

    (d) Voltage regulators
    (e) Communication subsystems
    (f) Active filters
    (g) Decade counter

2-2. *Medium-scale-integration* (MSI) is generally considered as containing a number of active components less than
    (a) 20
    (b) 50
    (c) 100
    (d) 500

2-3. The resistors contained in an LIC *chip* can be considered to have satisfactory tolerance within the following limits:
    (a) Within 10%
    (b) Within 50%
    (c) Within a ratio between resistors of ±10%

2-4. In fabricating a monolithic LIC, the following components are generally avoided for inclusion in the chip:
    (a) Resistors
    (b) High-value capacitors
    (c) Coils

2-5. An outstanding advantage of the transistors contained in an LIC array is
    (a) Good matching between individual transistors
    (b) High values of beta (or h_{fe})
    (c) High-power dissipation
    (d) Space saving

2-6. Biasing of the transistors in an LIC chip is generally
    (a) By self-contained circuitry
    (b) External resistors
    (c) Not necessary

2-7. Flexibility in the use of operational amplifiers is generally accomplished by
    (a) Controlled feedback in closed-loop operation
    (b) High open-loop operation
    (c) High current output operation

2-8. Freedom from oscillation in the direct-coupled OP AMP is ensured by the following:
    (a) No capacitors (internal or external) are required
    (b) The well-matched internal transistors
    (c) Frequency-compensation capacitors

2-9. The power-supply for OP AMPs is generally specified as
(a) Single +5-V supply
(b) Dual ±5-V supply
(c) Dual ±15-V supply

2-10. The noninverting input for OP AMPs is generally used
(a) To avoid an output change of polarity
(b) To provide positive feedback for an oscillator
(c) To provide a high input impedance

Chapter 3

Differential Amplifier Stage in Integrated-Circuit Design

3-1 GENERAL ELEMENTS IN INTEGRATED-CIRCUIT DESIGN

In the previous introductory discussion it was emphasized that the IC can be used as if it were a single active device, and that this ease of use provides a tremendous simplification—even for nontechnical personnel—in setting up a desired circuit. However, we need to know more than just the external connections if we are to have a good understanding of the basic functioning of the little "black box." By delving a bit further into the functional design elements of the internal arrangements of the IC, we can learn how to make more effective use of its many capabilities, while, equally important, being intelligently guided as to its inherent limitations.

The general problem of translating the design of a multistage transistor amplifier into an integrated-circuit form can be envisioned by the three blocks shown in Fig. 3.1(a). For the input stage of the IC, it has been found that the integrated version of a pair of transistors in the form of a *differential amplifier* gives the best answer for satisfying the critical input conditions, in order to produce stable bias, minimum offsets, and the ability to have simple provision for external negative feedback. Since this type of input stage has so great an effect in determining many characteristics of the resulting LIC, it will be discussed separately in this chapter. The intermediate stage often is a cascaded differential amplifier and the output stage an emitter follower.

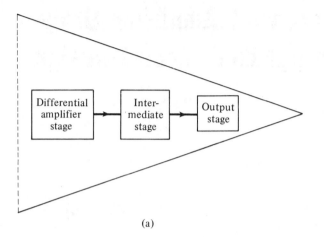

(a)

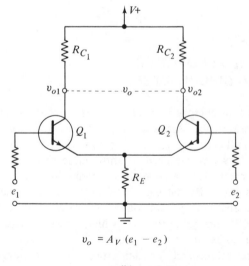

$$v_o = A_V (e_1 - e_2)$$

(b)

Figure 3.1 (a) Generalized blocks of an OP AMP; (b) simplified circuit of differential amplifier input stage of OP AMP; the output ($v_{o1} - v_{o2}$) is proportional to the difference of the inputs ($e_1 - e_2$).

3-2 CHOICE OF DIFFERENTIAL AMPLIFIER FOR INTEGRATED-CIRCUIT INPUT STAGE

In the most widely used forms the circuit arrangement of the linear IC should be able to provide a high order of amplification. While there are many versions of multistage ac amplifiers to accomplish the desired high gain, they necessarily include fairly large capacitors for *RC* coupling between stages—a condition that serves to rule out this circuit approach for integration. The circuit choice then narrows down to some form of multistage *direct-coupled amplification*, and this brings with it the attendant *problem of drift*, a characteristic that is inherent in dc amplifiers. This is so because unavoidable changes in operating conditions (such as changes in temperature, supply voltage, and so on) are accepted by a dc amplifier in the same way as it accepts the desired input signal, thus producing undesired output changes that together constitute the drift condition. Even small amounts of this undesired drift are important, since they will be magnified by the high gain of the subsequent stages. Consequently, the type of circuit that minimizes this drift, such as the differential-amplifier stage, is a preferred choice.

The *differential-amplifier stage* offers a good practical solution to the drift problem (except for the most stringent drift specifications, where a chopper-stabilized amplifier might be required). As shown in Fig. 3.1(b), it consists of a matched pair of transistors, where the drift-induced changes affecting one side of the pair tend to be canceled out by the corresponding changes on the other side of the pair. As a result of this balancing action, the differential-amplifier circuit is able to produce a substantially zero dc output under quiescent (no-signal) conditions. (This feature of "zero out for zero in" accounts for its popular use in the bridge type of electronic voltmeters.[1]) Because of this and other beneficial features, the circuit has been almost universally adopted in integrated circuits for the input stage (and often for intermediate stages as well). Variously designated as *differential-amplifier stage*, *emitter-coupled stage*, or *"long-tailed pair*," its advantageous features can be listed as follows:

1. It is *direct-coupled*, allowing both dc and ac amplification without requiring coupling capacitors.
2. It tends to *cancel drift conditions* that cause offset changes in the output, and simultaneously it can be made to be almost impervious to signals that are common to both inputs (such as hum pickup), because of its property of *common-mode rejection* (CMR).
3. Since its emitter resistance supplies internal negative feedback, the resulting amplifier characteristics show improvement in its *stability*, *wide-band response*, and *high input impedance*.

[1] Details of its use in solid-stage multimeters (such as the FET-VM) are given in S. Prensky, *Electronic Instrumentation*, 2nd ed., Prentice-Hall, Inc., Englewood Cliffs, N.J., 1971.

(It should also be noted that the disadvantage of the reduced gain, normally associated with negative feedback, is not a pressing problem in the case of integrated circuits, where it is relatively easy to fabricate additional transistors as required.)

With all these beneficial and practical characteristics, it is not surprising, therefore, to find the differential-amplifier circuit being used in the input of practically all integrated circuits as a *basic amplifier stage.*

3-3 CIRCUIT ACTION OF DIFFERENTIAL-AMPLIFIER STAGE

The performance of the differential-amplifier stage can be followed more clearly by starting with a functional circuit using a *matched pair of field-effect transistors,* as shown in Fig. 3.2. It is well to start with this circuit, because the operation of the unipolar FET requires no current in its input circuit (the conventional bipolar transistor, by contrast, requires input base-current flow as a necessary condition). Analysis of the FET circuit allows us to concentrate only on the output currents, making it much easier to grasp the relation between the output and input voltages in determining the voltage gain of the circuit. (This approach, with its concentration on the currents in the output circuit, will also be helpful in analyzing the bipolar transistor circuit, which is considered later.)

In arriving at an equivalent-circuit model for the emitter-coupled amplifier of Fig. 3.2(a), we can take advantage of the simplicity of the model[2] obtained by looking into the source resistor (R_{SC}), as shown in Fig. 3.2(b). Here the internal drain resistance (r_d) and load resistance ($R_D = R_L$), on each side of the circuit, are combined in a single equivalent resistor $(r_d + R_D)/(\mu + 1)$ and are fed by an equivalent generator $[\mu v_i/(\mu + 1)]$. We then make the simplifying assumptions that apply for good symmetry between the left and right sides; that is, $(r_{d1} = r_{d2} = r_d$, and $R_{D1} = R_{D2} = R_D)$. Also, we assume R_{SC} to be much larger than the equivalent series resistors (both of these assumptions are practical ones, as will be shown later).

Accordingly, as we trace the output signal current in Fig. 3.2(b) we can consider negligible current in R_{SC}, and find

$$i_{d1} = i_{d2} = i_d,$$

and the corresponding output voltages are

$$v_{o1} = -i_d R_D,$$

$$v_{o2} = +i_d R_D \quad \text{(equal and opposite to } v_{o1}).$$

[2] After J. Millman, *Microelectronics,* McGraw-Hill Book Company, New York, 1979, pp. 402–405.

FET parameters

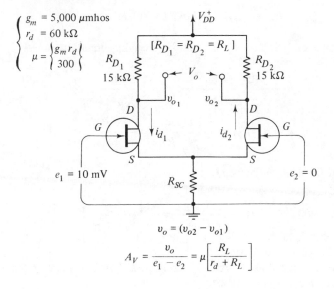

$$\begin{cases} g_m = 5{,}000\ \mu\text{mhos} \\ r_d = 60\ \text{k}\Omega \\ \mu = \begin{cases} g_m r_d \\ 300 \end{cases} \end{cases}$$

$$v_o = (v_{o2} - v_{o1})$$

$$A_V = \frac{v_o}{e_1 - e_2} = \mu\left[\frac{R_L}{r_d + R_L}\right]$$

(a)

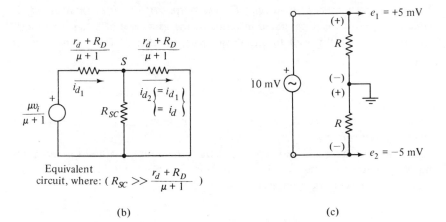

Equivalent
circuit, where: ($R_{SC} \gg \dfrac{r_d + R_D}{\mu + 1}$)

(b)   (c)

Figure 3.2 Functional circuit of FET differential-amplifier stage: (a) actual circuit; (b) equivalent circuit looking into source S; (c) positive and negative inputs giving same results as in (a).

Then, for the overall output voltage, $v_o = v_{o2} - v_{o1}$,

$$v_o = i_d R_D - (-i_d R_D) = 2(i_d R_D).$$

Finding the value of i_d and substituting in the preceding equation,

$$i_d = \frac{\mu v_i/(\mu + 1)}{2(r_d + R_D)/(\mu + 1)} \quad \text{or} \quad \frac{\mu v_i}{2(r_d + R_D)},$$

and
$$v_o = -2(i_d R_D) = \frac{-2\mu v_i(R_D)}{2(r_d + R_D)} = \frac{-\mu v_i R_D}{r_d + R_D}.$$

Therefore, for the differential gain (A_v),

$$A_v = \frac{v_o}{v_i} = \frac{-\mu R_D v_i}{(r_d + R_D)\, v_i} = -\mu \left[\frac{R_D}{r_d + R_D} \right],$$

which will be recognized as the *gain of a single FET, independent of the value of source resistance* (R_{SC}).

Output of General Differential-Amplifier Stage

Consequently, a *general expression for the output of the differential-amplifier circuit* $(v_{o2} - v_{o1})$ can be stated in terms of the gain (A_v) of a single element, and it applies equally well to any device pair, whether FETs or transistors, as follows:

$$v_o = v_{o2} - v_{o1} = A_v(e_1 - e_2).$$

Expressed in words, *the differential output is proportional to the difference of the two input voltages.*

Two examples will be given to illustrate this simple method for determining the approximate gain of the circuit; the first for the FET circuit and the other for a conventional (bipolar) transistor circuit.

Example 1
Using the FET Matched Pair: Single-ended Input to Differential Output

Using the numerical values given in Fig. 3.2(a) for the single-ended voltage input $(e_1 = 10 \text{ mV}, e_2 = 0)$, we first find the gain of a single stage (neglecting R_{SC}) as follows:

$$A_v = -\mu \, \frac{R_D}{r_d + R_D}$$

$$= -300 \, \frac{15 \text{ k}\Omega}{60 \text{ k}\Omega + 15 \text{ k}\Omega} = -300 \, \frac{15}{75} = -60.$$

Then, $v_o = v_{o2} - v_{o1} = -60(e_1 - e_2) = -60(10 \text{ mV} - 0)$, and $v_o = -600 \text{ mV}$ or -0.6 V as the differential output voltage, taken between the two drain resistors.

It may be noted that this is the same gain as would be obtained using the formula for the FET gain:

$$A_v = -g_m R_{\text{eff}}, \quad \text{where } R_{\text{eff}} = r_d \| R_D$$

With Differential Input

The input voltage of 10 mV could equally well be applied as a differential input, if desired, by arranging two equal resistors across the 10-mV generator, and connecting the center tap to ground, as in Fig. 3.2(c). Since the differential output will still be proportional to the difference of the input signals, we again have

$$v_o = v_{o2} - v_{o1} = A_v(e_1 - e_2)$$

$$= -60[5 \text{ mV} - (-5 \text{ mV})] \quad \text{or} \quad -60(10 \text{ mV})$$

$$= -600 \text{ mV},$$

the same differential output as before.

Single-Ended Output Taken from One Side to Ground

When it is desired to obtain an output voltage with respect to ground (rather than having the floating differential output), we take the output voltage from one drain resistor to ground, so that

$$v_o = v_{o1} \text{ (or } v_{o2}) \text{ to ground} = -i_d R_D = \frac{-A_v}{2} (e_1 - e_2)$$

$$= \frac{-60}{2} (10 \text{ mV})$$

$$= -300 \text{ mV};$$

thus the *single-ended output* (v_{o1} or v_{o2}) *will be half the differential output voltage.*

Conventional Transistor Circuit

The differential-amplifier circuit using conventional (or bipolar) transistors for the matched pair is by far the one most commonly used in integrated circuits. Even though a high input impedance can be obtained with the FET pair, the bipolar transistor pair is able to provide a *much higher gain* while still presenting a generally satisfactory input-impedance level. In analyzing the gain of the transistor version, shown in Fig. 3.3(a), we can again concentrate on the output circuit, as we did for the FET version. Even though additional considerations arise from the flow of signal current into each base, we find that the expression for the overall gain of the circuit again turns out to be the gain of a single transistor, independent of the emitter resistor. Thus, within a reasonable approximation, the differential output, taken between the two collectors, is again $v_o = v_{o2} - v_{o1} = A_v(e_1 - e_2)$, where A_v is the gain of a single transistor, as if it had no resistance in its emitter circuit. The simplified expression for A_v, shown in part (b) of the figure, makes use of the additional assumptions (usually valid) that h_{re} can be neglected and that $h_{oe}R_L$ (or in this case $h_{oe}R_C$) is much smaller than unity. Accordingly, we approximate the current gain:

$$A_i = \frac{h_{fe}}{1 + h_{oe}R_L} \approx h_{fe} \quad (\text{or } \beta),$$

and the differential voltage gain is

$$A_v = -h_{fe}\frac{R_L}{R_{in}} = -h_{fe}\left[\frac{R_L}{R_{gen} + h_{ie}}\right].$$

Example 2
Matched Pair of Bipolar Transistors

Using the numerical values given in Fig. 3.3(b), the approximate differential gain (A_v) is as follows:

$$A_v = \frac{-h_{fe}(R_L)}{R_{gen} + h_{ie}} = -100\left[\frac{7,500}{500 + 2,000}\right] = -100\left[\frac{75}{25}\right]$$

$$= -300,$$

which is five times as great as that for the FET pair.

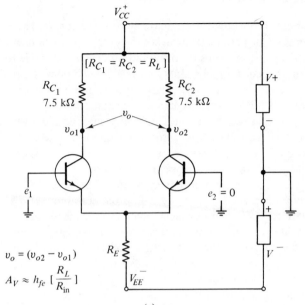

$$v_o = (v_{o2} - v_{o1})$$

$$A_V \approx h_{fe} \left[\frac{R_L}{R_{in}} \right]$$

(a)

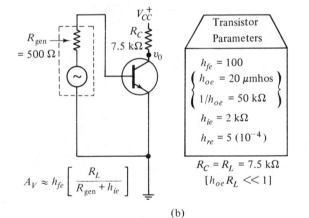

$$A_V \approx h_{fe} \left[\frac{R_L}{R_{gen} + h_{ie}} \right]$$

(b)

Figure 3.3 Bipolar (conventional) transistor version of a differential-amplifier stage: (a) actual circuit; (b) equivalent diagram for stage gain, showing gain to be essentially independent of value of emitter resistance R_E.

3-4 CONSTANT-CURRENT CIRCUIT

The desirability of making the common emitter resistor (R_E) as high as possible is based on a major advantage; it makes the output (v_{o1}) on one side of the pair more nearly equal and opposite to the output on the other side (v_{o2}). As a consequence of this equality, the ability to reject common-mode signals is greatly enhanced. This is expressed quantitatively as a desirably large *common-mode rejection ratio* (CMRR), which is discussed in a later section.

There is, however, a practical difficulty in choosing a very large resistor (R_E), since the large voltage drop across it would call for excessively large supply voltages to produce acceptable values of collector current. This difficulty is circumvented by adding a third transistor (Q_3) to the circuit, as shown in Fig. 3.4. This transistor acts as a constant-current source in place of R_E.

A constant-current source, by definition, corresponds to a source having an extremely large internal resistance. Such a source is obtained from transistor Q_3 by providing its base with an unvarying bias voltage, obtained in the usual case from the constant voltage across the diodes, which are forward biased from the $-V_{EE}$ supply. With the V_{BE} of Q_3 held constant in this way, the value of its collector current can be set to an appropriate value by its emitter resistor (R_3). The resulting constant collector current of Q_3 divides equally and establishes the quiescent collector currents of Q_1 and Q_2. (Looked at another way, tran-

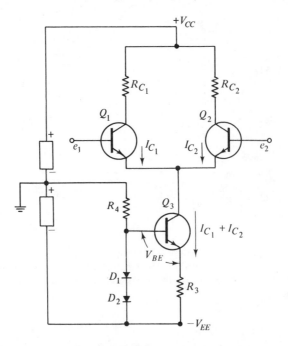

Figure 3.4 Constant-current circuit of Q_3 produces an effective very high resistance in place of emitter resistance (R_E) in Fig. 3.3 (a).

sistor Q_3 is being operated on the almost horizontal portion of its output characteristic curve, thus making its internal resistance equal to a very high resistance). In this manner, effective resistances equivalent to many megohms can be obtained without incurring excessive voltage drops in the actual circuit. Moreover, when this constant-current source is used in monolithic integrated circuits, an added advantage lies in the fact that the diodes used to bias Q_3 can easily be fabricated as diode-connected transistors on the same chip, thus supplying a good temperature-compensation match for Q_3. Since the stabilized Q_3 collector current determines the quiescent operating current, this method serves to reduce temperature drift for the whole circuit.

3-5 PRACTICAL CIRCUIT FOR SINGLE-ENDED STAGE

Because of the many beneficial characteristics of the differential-amplifier stage, and due to its comparative ease of flexible design, we find it used almost universally as the first stage of OP AMPS and of many other linear ICs. In many cases, a single-ended output from the stage is preferred, where the output is taken from only one of the two collectors, and hence the output may be referenced to ground (as opposed to the floating double-ended, or differential, output that is taken between each of the collectors). This form of single-ended output voltage is often a necessary practical consideration, even though the single-ended output has only half the gain of the double-ended case (as noted in Sec. 3-3). It is instructive, therefore, to examine a practical circuit design for such a single-ended case.

In the single-ended circuit of Fig. 3.5, we can omit the collector resistor R_{C1} (formerly used as collector resistor for Q_1 of Fig. 3.4), since no output will be taken from it; this results in no significant change in the total current ($I_{C3} = I_{C1} + I_{C2}$), since I_{C3} is dictated by the constant-current source (Q_3). For illustrative purposes, we will use one half of the dual differential amplifier in type NE 511 (*Signetics*).[3] The following values are used:

The total current ($I_{C3} = I_{C1} + I_{C2}$) = 3 mA, with a practically equal division of 1.5 mA for I_{C1} and I_{C2}; the input resistance (R_{in}) is approximately 5,000 Ω, determined primarily by the value of I_{C3}. The quiescent voltage, collector-to-ground (V_{C2}) for Q_2, is about 6 V, half of the $V+$ supply, and is obtained, by using $R_{C2} = 4$ kΩ, from

$$V_{CE} = (V+) - (I_{C2}R_{C2}) = 12 \text{ V} - (1.5 \text{ mA} \times 4 \text{ k}\Omega) = 6 \text{ V}.$$

The voltage amplification (A_v) for this single-ended stage can be calculated from the transistor transconductance ($g_m = 25$ millisiemens), as obtained from

---

[3]Adapted from *Signetics Analog Manual*, which also includes a thorough discussion of the design equations involved.

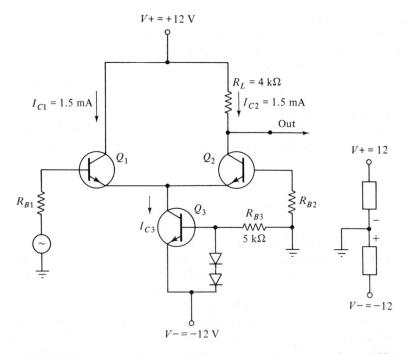

Figure 3.5 Practical example for obtaining single-ended output from differential amplifier.

the 511 data sheet, and the load resistance (R_L) by

$$A_v = -g_m R_L = -25(10^{-3}) \times 4(10^3) = -100.$$

The design values in this example are fairly typical, using a dual supply of ± 12 V to provide a voltage gain of -100, with an output-voltage swing of about ± 10 V; the values may be summarized as follows:

$$R_L = 4 \text{ k}\Omega,$$

$$R_{B3} \text{ (to provide stable constant current)} = 5 \text{ k}\Omega$$

$$I_{C3} = I_{C1} + I_{C2} = (1.5 + 1.5) \text{ mA} = 3 \text{ mA}.$$

The basic differential-amplifier stage (whether its output is single or double ended) is used in a variety of forms in linear ICs to accomplish specific goals. In the case of OP AMPS, for example, *increased open-loop gain* is generally provided by at least two stages of the basic diff-amp type of amplifier. For increased *common-mode rejection ratio* (CMRR), the first input stage generally has a double-ended output to achieve good rejection, as discussed in the next section. The double-ended output form also is used for obtaining a balanced *quiescent condition of zero output for zero input*, as required in dc amplification. Obtaining an *increased input resistance*, by use of Darlington-connected transistors, is also discussed in a later section.

3-6 COMMON-MODE REJECTION RATIO

A most important characteristic required in the first stage of a high-gain amplifier is its ability to suppress undesired disturbances that might be amplified along with the desired signal. The matched pair of transistors in the differential-amplifier stage has this inherent capability, based upon the fact that unwanted signals from an external source (such as hum pickup) would appear as common to both input bases, and as such, would produce equal output voltages whose difference in the overall output would theoretically be zero.

The practical effectiveness of this rejection, however, depends upon how closely equal the currents in the left and right sides of the circuit can be held. Referring to Fig. 3.2(b), it can be seen that the equality of those currents (i_{d1} and i_{d2}) depends on realizing the assumption of a sufficiently large common resistor (R_{SC} in the FET circuit and R_E in the transistor circuit). We have seen that the strategy of using a third transistor (as a constant-current source) contributes greatly to this desired condition. We can evaluate the extent of the common-mode rejection by use of the ratio comparing the desired differential amplification (A_d) with the undesired amplification of the common-mode signals (A_{cm}). (Note that A_d as used here corresponds to the voltage gain previously designated as A_v). A figure of merit for the common-mode rejection ratio (CMRR) is given as

$$\text{CMRR} = \frac{A_d}{A_{cm}}.$$

This ratio will depend upon how large a resistance is presented by the collector of Q_3 at the junction of the common emitters; its equivalent value is designated R_{cm}. The resulting CMRR can be expressed as the number of times that A_d is greater than A_{cm}, or, as is usually done, by the logarithmic ratio in decibels, as illustrated in the following example:

Example 3
Determining CMRR

We shall assume that measurements in the circuit of Fig. 3.4 give the following results: when tested for differential output (v_o) for single-ended input (e_{in}),

$$\text{input } (e_{in}) = e_1 - e_2 = 1 \text{ mV} - 0 = 1 \text{ mV};$$

$$\text{output (diff)} = e_o = v_{o2} - v_{o1} = 200 \text{ mV}.$$

Thus, *differential gain* $A_d = v_o/e_{in} = 200 \text{ mV}/1 \text{ mV} = 200$.

Also, when both input terminals are tied together, assume that a *common-mode* signal of 10 mV (v_{cm}) is applied to the common input, and a common-mode output of 2 mV is obtained:

$$\text{common mode gain } A_{cm} = v_{o(cm)}/v_{cm} = 2 \text{ mV}/10 \text{ mV} = 0.2,$$

where v_{cm} = 10 mV is unwanted pickup, say by both unshielded input leads, resulting in $v_{o(cm)}$ = 2 mV from imperfect cancellation. Then

$$\text{CMRR} = A_d/A_{cm} = 200/0.2 = 1,000 \quad (\text{or } 60 \text{ dB}).[4]$$

Note: The effectiveness of this fairly conservative value for the CMRR may be noted from this example; in spite of the fact that the common-mode signal is 10 times as great as the differential signal, the differential output of 200 mV contains only a negligible amount of common-mode output (2 mV, or about 1 percent). More typical values of 90 to 100 dB for the CMRR are obtainable, thus greatly reducing this percentage of common-mode error in well-designed differential-amplifier stages.

3-7 INPUT-STAGE VARIATIONS IN PRACTICAL INTEGRATED CIRCUITS

In examining schematic diagrams of commercial ICs, we often find that the input to each base of the matched pair of transistors is applied to a Darlington-connected arrangement for each side of the pair, as shown in Fig. 3.6. This refinement (often found in operational amplifiers) is designed to achieve a higher input impedance and so to reduce the quiescent bias currents, especially for low-level signals. Other versions might be found using FETs for the matched pair to achieve an even greater value of input impedance. Various other stratagems using additional transistors are employed in practical linear ICs to achieve certain specific characteristics, and these are detailed in later chapters.

Summarizing the Differential-Amplifier Stage

The foregoing discussion may be summarized by saying that the *differential-amplifier circuit* has become practically a *standard for the input stage of an*

---

[4] For conversion to decibels, see Appendix I.

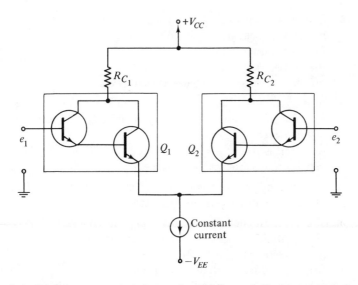

Figure 3.6   Darlington-connected transistors, Q_1 and Q_2, serve to reduce the amount of bias current required and, therefore, to increase the input resistance of the circuit. The constant current circuit connected to the emitters is represented by a current source symbol.

integrated circuit, because of the greatly improved overall characteristics that it offers as a stable amplifier; its performance generally exceeds most other configurations, even when using discrete components. The benefits obtained by the use of three integrated transistors for this input stage may be listed as improved values of *high input impedance* and very good *suppression of drift and undesired signals.* Another feature of integrated circuits that may be noted here is the free use of a large number of transistors that would usually not be practical in a discrete circuit. Many other examples of this generous use of transistors fabricated in a single chip for specific objectives will be seen in subsequent discussions of typical integrated circuits.

QUESTIONS

    3-1. A direct-coupled (dc) amplifier can be used to amplify
        (a) Only dc signals
        (b) Only ac signals
        (c) Both types of signals

    3-2. A major problem with the direct-coupled amplifier, which is not as troublesome in capacitor-coupled (ac) amplifiers, is
        (a) High gain
        (b) Frequency response
        (c) Drift

3-3. The voltage gain of a differential amplifier stage using two transistors in a balanced circuit is equal to the gain
  (a) Of a single stage
  (b) Of two stages
  (c) Between that of a single stage and two stages

3-4. In the functional circuit of a differential amplifier stage (Fig. 3.1), if the value of the emitter resistor R_E is increased, the voltage gain is
  (a) Increased
  (b) Decreased
  (c) Remains substantially the same

3-5. In question 3-4, what is the effect on the voltage gain if the values of the collector resistors are increased, instead of R_E?

3-6. The practical method for increasing the common-mode rejection ratio (CMRR) of a differential-amplifier stage is to
  (a) Increase the value of R_E
  (b) Substitute a constant-current source for R_E
  (c) Reduce the value of R_E

3-7. A transistor connected as a current source provides the effect of
  (a) A very high input Z
  (b) A very low input Z
  (c) A very low output Z
  (d) A very high output Z

3-8. In the constant-current version of the differential amplifier (Fig. 3.4), the constant current in Q_1 and Q_2 is caused by
  (a) The constant values of V_{CC} and V_{EE}
  (b) The constant current through diode's D_1 and D_2
  (c) The constant bias on Q_3

3-9. In the same constant-current circuit (Fig. 3.4), temperature compensation for Q_3 is provided by
  (a) V_{BE} of Q_1 and Q_2
  (b) Voltage drops across D_1 and D_2
  (c) Voltage drop across R_4

PROBLEMS

3-1. In Example 1 (Sec. 3-3), find voltage amplification (A_v) by using the relation
  (a) $A_v = -g_m R_{eff}$, where $R_{eff} = R_D \| r_d$

  Note: The g_m of the FET is often represented by y_{fs}.

    (b) Compare your result with the answer given in Example 1 in the text, using the relation

$$A_v = -\mu \left[\frac{R_D}{r_d + R_D} \right]$$

3-2. In Example 1, as given above, when the value of the internal drain resistance (r_d) is sufficiently large $(r_d \gg R_D)$, we can use the approximate formula

$$A_v \approx -g_m R_D \quad (\text{or} \approx -y_{fs} R_D)$$

Find the percentage of error in using the approximate formula $A_v \approx -g_m R_D$, in Example 1
(a) When $r_d = 60$ kΩ, $R_D = 15$ kΩ
(b) When $r_d = 390$ kΩ, $R_D = 15$ kΩ

3-3. In Example 3 (Sec. 3-6), the common-mode gain is given as $A_{cm} = 0.2$ in a circuit where the differential gain is $A_d = 200$, resulting in CMRR = 1,000 (or 60 dB). Find the new common-mode rejection ratio (CMRR) (in magnitude and in decibels) if better matching of the transistors results in a common-mode gain $(A_{cm} = 0.1)$, while A_d (at 200) remains the same.

3-4. For expressing voltage gains in decibels, it is well to memorize the values which are left blank (and are to be filled in) in the following tabulation. (These values are printed in **boldface** in the Decibel Table of Appendix I.)

Numerical gain ratio (X)	No. of dB
2X	
10X	
100X	
	3 dB
	6 dB
	20 dB
	40 dB

3-5. Using the boldface values in the Decibel Table of Appendix 1, find
(a) The number of decibels for
    (a$_1$) Voltage gain (VG) = 200
    (a$_2$)   = 400
    (a$_3$)   = 2,000

(b) The numerical voltage gain ($\times$) for

(b$_1$) 12 dB

(b$_2$) 23 dB

(b$_3$) 52 dB

3-6. Find the CMRR in decibels, when the CMRR ($= A_d/A_{cm}$) is given as follows:

(a) When A_d/A_{cm} = 100,000

(b) When A_d/A_{cm} = 20,000

3-7. In designing the differential-amplifier stage of Fig. 3.3, find the new value for voltage amplification (A_v), when all values stay the same, but

(a) Change h_{fe} from 100 to 150

(b) Change $R_C = R_L$ from 7.5 to 15 kΩ, with h_{fe} = 150

(c) Change R_{gen} from 500 Ω to 3 kΩ, with h_{fe} = 150 and $R_C = R_L$ = 15 kΩ

Chapter 4

Operational Amplifier Characteristics

4-1 BASIC REQUIREMENTS FOR THE OPERATIONAL AMPLIFIER

The most prominent form of linear integrated circuit, by far, is the *operational amplifier*. In the amplification process it is generally important for the output to be a closely faithful reproduction of the input, and so *linear operation* is one of the major characteristics of this integrated circuit, accounting for the designation of "linear" in the LIC category. [The prominence of the operational amplifier types in this nondigital category explains, perhaps, the stretching of the designation of linear IC to include even some types having a nonlinear output, such as comparators and regulators, in order to distinguish them from the digital (DIC) category.] Be that as it may, we can start with linearity of operation as one of the essential characteristics of the operational amplifier.

The OP AMP derives its highly attractive properties as an active device from its basic form of a very high gain amplifier, which also has provision for external feedback. In this form the output of the amplifier can be made to depend primarily on *externally connected passive elements*, so *its amplifying performance is virtually independent of its internal parameters*. To accomplish this highly flexible form of amplifier, there are three main requirements for the internal circuitry:

1. The *open-loop gain A_{VOL} of the amplifier must be very high* (preferably well over 10,000 times, or 80 dB).

55

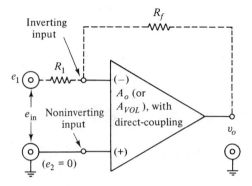

Figure 4.1 Symbol of OP AMP used as an inverting amplifier, emphasizing three main requirements: (1) very high open-loop gain, A_o or A_{VOL}; (2) direct-coupled stages; (3) external resistor (R_f) for negative feedback from output to inverting input (labeled minus).

2. The *multiple stages must be direct coupled* (this allows both dc and ac amplification).
3. There must be provision for obtaining *an inverted output from which negative feedback can be obtained by a single external resistor* connected from the output to the input terminal of the amplifier.

These three basic requirements are shown schematically in Fig. 4.1, in which the triangle represents the large open-loop gain (A_{VOL}, abbreviated here as A_o); the two inputs (for dc or ac amplification) are labeled (−) for the inverting function and (+) for the noninverting function; also the feedback resistor (R_f) is shown connected externally between the output terminal and the inverting input (−) terminal to provide negative feedback. Under these basic conditions, the gain of the operational amplifier with feedback (A_f) (or, as usually stated, the closed-loop gain, A_{VCL}) used in an *inverting* mode can be expressed by the simple relation

$$A_f = A_{VCL} = -\frac{R_f}{R_1} .$$

The analysis for tracing this relation for the gain with feedback (or closed-loop gain) is discussed in the next section.

4-2 ANALYSIS AND PRACTICAL EXAMPLE OF OPERATIONAL-AMPLIFIER ACTION

When the requirement for a very large open-loop gain is satisfied, we can analyze the circuit action of the operational amplifier in a simplified manner to readily

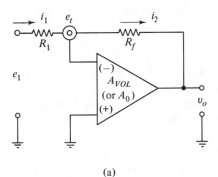

(a)

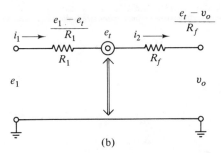

(b)

Figure 4.2 Concept of virtual ground: (a) circuit; (b) virtual ground shown as double-ended arrow.

yield the relationship for the amplifier gain with feedback. Referring to the generalized circuit of Fig. 4.2(a), and considering the situation at the inverting input (−) of the amplifier, the voltage at that terminal is labeled e_t. The output voltage (v_o) can then be considered as the voltage resulting from the voltage at the terminal (e_t) multiplied by the open-loop gain, or

$$v_o = A_o(e_t)$$

and

$$e_t = \frac{v_o}{A_o}.$$

From this it can be seen that as open-loop gain (A_o) becomes larger, e_t becomes progressively smaller, so that e_t approaches zero as A_o approaches infinity. It follows, then, that in a practical amplifier with A_o sufficiently large (> 10,000, and usually much larger) e_t can be considered as practically zero. (This remains true for all values of input voltage within the linear range of the amplifier, since larger input voltages result in correspondingly greater values of the feed-back voltage.)

Virtual Ground

The situation where the voltage at the amplifier terminal (e_t) is substantially zero is known as a *virtual ground*, signifying a point that is *substantially at ground level, but not physically connected to ground*. Since we have both points $(e_t$ and ground) at practically the same potential, we may assume that practically no current will flow between them. Any current i_1 entering terminal e_t may be considered as flowing directly to the output terminal, as current i_2.

By using this concept of a virtual ground[1] [shown in Fig. 4.2(b) by a heavy double-ended arrow], we can readily arrive at an expression for the gain of the circuit. Here we can equate the current i_1 and i_2 as follows:

$$i_1 = \frac{e_1 - e_t}{R_1} = i_2 = \frac{e_t - v_o}{R_f}.$$

With e_t as substantially zero,

$$\frac{e_1}{R_1} = -\frac{v_o}{R_f}.$$

Thus the gain of the circuit with feedback is

$$A_f = A_{VCL} = \frac{v_o}{e_1} = -\frac{R_f}{R_1}$$

with the negative sign indicating an inverted output.

Example 1
Type 741 OP AMP as Microphone Amplifier

A simple but practical example, to illustrate how easy it is to apply the OP AMP for general-purpose amplification, is given in the inverting amplifier circuit of Fig. 4.3(a), which uses the OP AMP for a microphone amplifier. Suppose we wish to use a handy loudspeaker as a microphone (as is done in intercom sets), and find that the feeble output voltage from the voice coil of our makeshift micro-

[1]G. J. Deboo and C. N. Burrous, *Integrated Circuits and Semiconductor Devices*, 2nd ed., McGraw-Hill Book Company, New York, 1977.

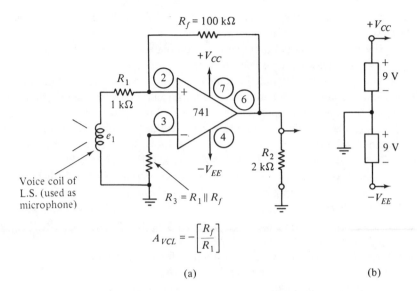

Figure 4.3   Simple example of OP AMP used as inverting amplifier (gain $A_v =$ -100): (a) circuit; (b) power supply connections.

phone needs to be amplified to produce a respectable output voltage. By using an OP AMP for this amplification instead of discrete transistors, the task of building the little amplifier is considerably simplified.

For the OP AMP, we will use the very popular type 741 in its eight-pin mini-DIP version (mini-dual-in-line package). This type is readily available as a type μA741 (Fairchild), or as an alternate-source type 741, having a different prefix from many other manufacturers. It will be noted that this simple amplifier circuit requires only five of the eight available pins, that is, two input, two power supply, and one output terminal, with the ground connection made externally at the mid-connection of the two 9-V batteries, as shown in Fig. 4.3(b). We will leave the construction details regarding the IC socket and the breadboarding of the circuit until the discussion in Chapter 6 and, for now, turn our attention to the three resistors needed as external components to achieve a desired voltage gain of, say, 100 times.

We determine the amount of gain for this inverting mode very simply by the ratio of the two resistors, R_f/R_1, where R_f is the feedback resistor and R_1 the input resistor to the inverting terminal, labeled $(-)$. By choosing a convenient value for R_1 of 1 kΩ, we get the desired gain of 100 by simply making R_f 100 times as great, or $R_f = 100$ kΩ. (Note that, since only voltage gain, rather than power gain, is required, we need not bother here with matching the 45-Ω internal resistance of the source, as long as we make the input resistor R_1 at least

20 times larger than the source resistance of 45 Ω so that most of the microphone voltage reaches the input terminal).

The noninverting (+) input terminal is connected to R_3, which is returned to ground. The load R_L could be a headset of 2 kΩ or a 2-kΩ resistor as shown, from which the output would be taken to a power amplifier. Connecting the dual supply from the two transistor batteries completes the connections and produces the required output signal, amplified about 100 times. Resistor R_3 is used to equalize the bias currents flowing into the two input transistors of the differential amplifier and thereby minimize the offset voltage. The value of $R_3 = R_1 \| R_f$. Because $R_f = 100$ kΩ $>> R_1 = 1$ kΩ, $R_3 \simeq R_1 = 1$ kΩ.

While this particular example of an OP-AMP application is fairly straightforward, since it only requires the limited bandwidth needed for reproducing audio frequencies, it is sufficiently typical to bring out the outstanding simplicity of using the OP AMP; we can immediately appreciate the fact that *the* OP AMP *dispenses entirely with the need for complicated calculations*, either for achieving the desired gain or, for that matter, for any calculations for the proper bias, as would be the case with discrete transistors.

Moreover, even though other factors will enter later for special applications (such as, for example, minimum dc offset), we will find that these special characteristics can also be handled in a comparatively forthright manner by selecting the proper OP AMP and adding appropriate external components as specified for each such application. In this simple application, we also can see an additional advantage in the fact that no coupling capacitors at all are needed in the circuit, thus making it suitable for dc as well as ac applications.

This introductory example serves to emphasize *simplicity of operation* as one of the features of the OP AMP, while later discussions will point out the *wide versatility* as another important feature.

In summary, this analysis demonstrates that the closed-loop gain of the practical inverting amplifier can be closely approximated as the *ratio of two external resistances*, based upon the action of an operational amplifier having a very large open-loop gain, and also arranged for negative feedback by means of an external resistor (R_f).

4-3 FLEXIBLE USE OF OPERATIONAL-AMPLIFIER CIRCUITS

Before entering upon the details of operational-amplifier characteristics, it is well to quickly glance at the variety of useful circuits that can be assembled quite easily by using an inexpensive general-purpose version of this versatile IC, considered as a single device, along with a few external components.

An introductory listing of 12 useful OP AMP circuits may be roughly divided into three convenient categories: general-purpose amplification (both as dc or ac amplifiers), analog computer elements, and miscellaneous uses, as follows:

A. *General-purpose amplification*
   1. Inverting amplifier (for low-impedance signals).
   2. Noninverting amplifier (for high-impedance signals).
   3. Unity-gain amplifier (as voltage follower).
B. *Analog-computer elements*
   4. Summing amplifier (including a subtracting function).
   5. Integrator (for solution of differential equations).
   6. Differentiator (for pulse peaking).
C. *Miscellaneous uses*
   7. Linear rectifier (especially useful for low-level ac signals).
   8. Active filters (low-pass, high-pass, and all-pass phase-shifting circuits).
   9. Logarithmic amplifier (for wide dynamic range).
   10. Sine-wave oscillator circuits.
   11. Multivibrator circuits (including astable, monostable, and bistable circuits).
   12. Function generator (including square-wave, triangular-wave, and ramp signals and pulses).

The circuit applications listed are surely ample evidence of the flexibility of this OP AMP branch of the LIC field; moreover, it should also be borne in mind that new uses continue to be added for many particular applications. Within space limitations, many of these circuits will be described in Chapter 5; other circuit developments can be found in the references cited for further study, which will be found within the chapter and at the end of Chapter 12.

4-4 DATA SHEET: TYPE 741 op-amp CHARACTERISTICS

See pages 62, 63 and 64 for data sheets.

The chart showing *electrical characteristics* on the data sheet for the general purpose type 741C applies to the *commercial grade* (operating temperature range $0°$ to $70°C$). The range for the *military grade*, as noted on the first page of the data sheet, is greater ($-55°$ to $125°C$). Also available from Fairchild (and other manufacturers) is an improved type (741A/E), having tighter specifications for special uses, particularly with respect to lower values of input bias current and offset voltage. However, the specifications shown for type 741C

µA741
FREQUENCY-COMPENSATED OPERATIONAL AMPLIFIER
FAIRCHILD LINEAR INTEGRATED CIRCUITS

GENERAL DESCRIPTION — The µA741 is a high performance monolithic Operational Amplifier constructed using the Fairchild Planar* epitaxial process. It is intended for a wide range of analog applications. High common mode voltage range and absence of "latch-up" tendencies make the µA741 ideal for use as a voltage follower. The high gain and wide range of operating voltage provides superior performance in integrator, summing amplifier, and general feedback applications.

- NO FREQUENCY COMPENSATION REQUIRED
- SHORT CIRCUIT PROTECTION
- OFFSET VOLTAGE NULL CAPABILITY
- LARGE COMMON-MODE AND DIFFERENTIAL VOLTAGE RANGES
- LOW POWER CONSUMPTION
- NO LATCH UP

ABSOLUTE MAXIMUM RATINGS

Supply Voltage	
Military (741)	±22 V
Commercial (741C)	±18 V
Internal Power Dissipation (Note 1)	
Metal Can	500 mW
DIP	670 mW
Mini DIP	310 mW
Flatpak	570 mW
Differential Input Voltage	±30 V
Input Voltage (Note 2)	±15 V
Storage Temperature Range	
Metal Can, DIP, and Flatpak	−65°C to +150°C
Mini DIP	−55°C to +125°C
Operating Temperature Range	
Military (741)	−55°C to +125°C
Commercial (741C)	0°C to +70°C
Lead Temperature (Soldering)	
Metal Can, DIP, and Flatpak (60 seconds)	300°C
Mini DIP (10 seconds)	260°C
Output Short Circuit Duration (Note 3)	Indefinite

EQUIVALENT CIRCUIT

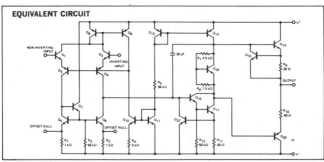

Notes on following pages.

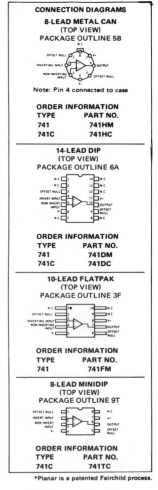

CONNECTION DIAGRAMS

8-LEAD METAL CAN
(TOP VIEW)
PACKAGE OUTLINE 5B

Note: Pin 4 connected to case

ORDER INFORMATION

TYPE	PART NO.
741	741HM
741C	741HC

14-LEAD DIP
(TOP VIEW)
PACKAGE OUTLINE 6A

ORDER INFORMATION

TYPE	PART NO.
741	741DM
741C	741DC

10-LEAD FLATPAK
(TOP VIEW)
PACKAGE OUTLINE 3F

ORDER INFORMATION

TYPE	PART NO.
741	741FM

8-LEAD MINIDIP
(TOP VIEW)
PACKAGE OUTLINE 9T

ORDER INFORMATION

TYPE	PART NO.
741C	741TC

*Planar is a patented Fairchild process.

(Fairchild Camera and Instrument Corp.)

FAIRCHILD LINEAR INTEGRATED CIRCUITS • μA741C

741C

ELECTRICAL CHARACTERISTICS (V_S = ±15 V, T_A = 25°C unless otherwise specified)

PARAMETERS (see definitions)	CONDITIONS		MIN.	TYP.	MAX.	UNITS
Input Offset Voltage	$R_S \leqslant$ 10 kΩ			2.0	6.0	mV
Input Offset Current				20	200	nA
Input Bias Current				80	500	nA
Input Resistance			0.3	2.0		MΩ
Input Capacitance				1.4		pF
Offset Voltage Adjustment Range				±15		mV
Input Voltage Range			±12	±13		V
Common Mode Rejection Ratio	$R_S \leqslant$ 10 kΩ		70	90		dB
Supply Voltage Rejection Ratio	$R_S \leqslant$ 10 kΩ			30	150	μV/V
Large Signal Voltage Gain	$R_L \geqslant$ 2 kΩ, V_{OUT} = ±10 V		20,000	200,000		
Output Voltage Swing	$R_L \geqslant$ 10 kΩ		±12	±14		V
	$R_L \geqslant$ 2 kΩ		±10	±13		V
Output Resistance				75		Ω
Output Short Circuit Current				25		mA
Supply Current				1.7	2.8	mA
Power Consumption				50	85	mW
Transient Response (Unity Gain)	Risetime	V_{IN} = 20 mV, R_L = 2 kΩ, $C_L \leqslant$ 100 pF		0.3		μs
	Overshoot			5.0		%
Slew Rate	$R_L \geqslant$ 2 kΩ			0.5		V/μs

The following specifications apply for 0°C $\leqslant T_A \leqslant$ +70°C:

Input Offset Voltage					7.5	mV
Input Offset Current					300	nA
Input Bias Current					800	nA
Large Signal Voltage Gain	$R_L \geqslant$ 2 kΩ, V_{OUT} = ±10 V		15,000			
Output Voltage Swing	$R_L \geqslant$ 2 kΩ		±10	±13		V

TYPICAL PERFORMANCE CURVES FOR 741C

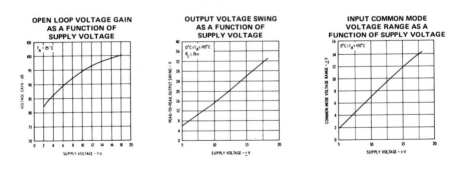

OPEN LOOP VOLTAGE GAIN AS A FUNCTION OF SUPPLY VOLTAGE

OUTPUT VOLTAGE SWING AS A FUNCTION OF SUPPLY VOLTAGE

INPUT COMMON MODE VOLTAGE RANGE AS A FUNCTION OF SUPPLY VOLTAGE

NOTES:
1. Rating applies to ambient temperatures up to 70°C. Above 70°C ambient derate linearly at 6.3 mW/°C for the Metal Can, 8.3 mW/°C for the DIP, 5.6 mW/°C for the Mini DIP and 7.1 mW/°C for the Flatpak.
2. For supply voltages less than ±15 V, the absolute maximum input voltage is equal to the supply voltage.
3. Short circuit may be to ground or either supply. Rating applies to +125°C case temperature or 75°C ambient temperature.

(Fairchild Camera and Instrument Corp.)

FAIRCHILD LINEAR INTEGRATED CIRCUITS • μA741

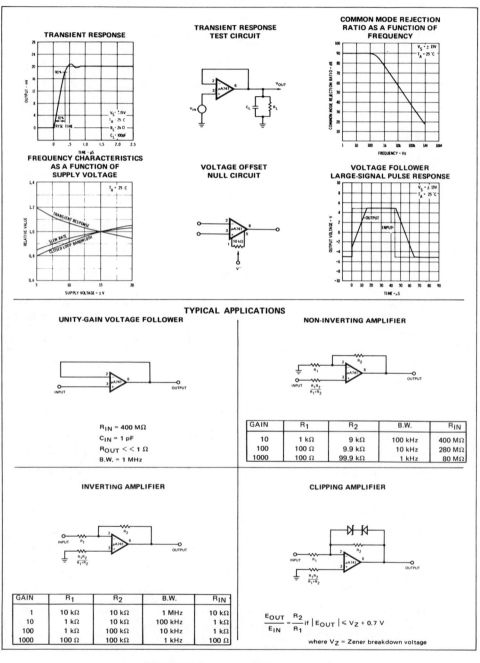

(Fairchild Camera and Instrument Corp.)

will be found to be amply adequate for most general-purpose uses in ac amplification, and also even for cases of dc amplification (where the gain is fixed), by using the voltage offset null circuit to reduce the offset to zero, as shown on the third sheet of data.

The complete data bulletin generally includes a number of graphs showing the relation of OP-AMP performance to various conditions (temperature, power-supply voltage, and the like). An especially important graph, for open-loop fre-

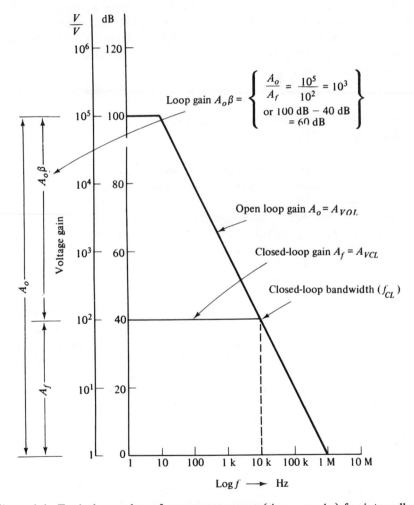

Figure 4.4 Typical open-loop frequency-response (A_{VOL} or A_o) for internally compensated OP AMP, showing relation of closed-loop gain (A_{VCL} or A_f): $A_f = \dfrac{A_o}{1 + A_o \beta} \approx \dfrac{A_o(\text{open-loop gain})}{A_o \beta(\text{loop gain})}$.

quency response versus voltage gain, is reproduced in Fig. 4.4. It enables us to tell how the closed-loop bandwidth varies for different values of voltage gain, as discussed in the next section.

At first glance the data sheet of an operational amplifier shows a large number of characteristics that can be quite bewildering to anyone who is primarily interested in assembling a simple amplification function. But the numerous specifications, however desirable in some special instances, are by no means equally necessary for general-purpose use. For the sake of simplicity, then, it is well to first concentrate on just a few *major characteristics*, which are defined and illustrated in the next section; more refined characteristics are left for a succeeding section, in which more exacting requirements will be discussed. (Additional notes on interpretating this and other data sheets are given in Section 4-7.)

4-5 MAJOR OPERATIONAL-AMPLIFIER CHARACTERISTICS

A first consideration in evaluating an OP AMP for general-purpose use is its *gain–bandwidth product* for voltage amplification. Since the product of gain times bandwidth $(G \times BW)$ is generally a constant (around which gain can be traded off for bandwidth), this characteristic for each OP AMP is usually shown on the data sheet by means of a *semilog plot of open-loop gain (A_{VOL}) in decibels against log frequency (f)*.

As shown in Fig. 4.4, one obtains the approximate bandwidth for any desired closed-loop gain by simply drawing a horizontal line from the desired value of gain to intersect the slope of the gain roll-off. The slope shown in Fig. 4.4 is the 6-dB/octave (or 20-dB/decade) slope that is typical of an internally compensated OP AMP (such as the popular 741 type and similar types). For the gain–bandwidth product of such an OP AMP, the graph shows the closed-loop gain of 100 times (40 dB) intersecting the slope at a cutoff frequency $(f_{c(-3 \text{ dB})})$ of 10 kHz, giving a $G \times BW$ *product of 1 MHz*. It will be noted that this agrees with the unity gain (0 dB) value of 1 MHz, giving the same product (1 × 1 MHz).

A relatively modest extension of the gain–bandwidth product is a feature of another popular OP AMP (the 748 type among others) that, for example, uses a single external compensating capacitor of 3 pF to extend the -3-dB bandwidth from 10 to 100 kHz at the same gain of 100. With this simple change of compensating capacitor, *the $G \times BW$ product is increased to 10 MHz*.

Since the graph for open-loop gain $(A_{VOL}$ versus $f)$ applies only for small-signal operation, a specification is usually included for *large-signal operation*. Here the value of gain is stated for the case where relatively full output swing is required ($V_{out} = \pm10$ V). In the popular examples cited thus far, this large-signal gain is generally between 15,000 and 25,000 times.

In the customary specification data on *electrical characteristics*, the preceding major characteristics are summarized as follows:

1. *Open-loop gain* (A_{VOL} *or* A_o). The voltage-amplification ratio of output to input voltage without external feedback illustrated in Fig. 4.1, with R_f disconnected:

$$A_{VOL} = v_o/e_{in}, \quad \text{where } R_f = \infty.$$

2. *The G × BW product for small-signal operation.* Obtained from the plot of open-loop frequency response, as explained previously; further details for wide-band operation are discussed later under the topic of frequency-compensation methods.

3. *Large-signal voltage gain* (v_o/e_{in}). The ratio of the maximum output-voltage swing to the change in input voltage required to drive the output from zero to this voltage.

4. *Input resistance* (R_{in}). Associated with the values of available gain is the input resistance that the OP AMP presents to the signal source. This depends, among other things, on the *input bias current* (at zero output). In effect, R_{in} determines the portion of the signal voltage remaining effective at the input after deducting any significant voltage drop across the internal resistance of the source.

4-6 COMPARISON CHART OF TYPICAL VALUES OF OP-AMP CHARACTERISTICS

A comparison of some *typical values* for the characteristics of fairly standardized OP AMPS is also very helpful in obtaining a clear view of the major features among the many refined specifications that proliferate in this highly competitive field.

The *"standardized"* OP AMPS have evolved from what is thought of as the first-generation types (such as the 709 and 741 types), through second-generation types (such as the 101A and 1556 types), and on into third-generation types that are known as *improved-specification types*. These tend to approach the specialized precision-instrumentation types considered in Chapter 11, but still remain in the general-purpose class. Some of these improved-specification types will be exemplified in Section 6-9 in connection with testing for such characteristics.

TABLE 4-1

Some Typical Operating-Amplifier Comparisons[a]

	Device Type[b]									
25°C	*709*	*709A*	*741*	*748*	*101A*	*777*	*770*	*660*	*108*	*108A*
Input offset voltage (mV)	<5	<2	<5	<5	<2	<2	<4	<3	<2	<0.5
Input offset current (nA)	<200	<50	<30	<200	<10	<3	<2	<5	<0.2	<0.2
Input bias current (nA)	<500	<200	<200	<500	<75	<25	<15	<15	<2	<2
Input resistance (MΩ)	>0.040	>0.085	>1	>0.5	>1.5	>2	100 typ	>4	>30	>30
Large-signal voltage gain[c]	45,000	45,000	>50,000	>50,000	>50,000	>50,000	>50,000	>25,000 $R_L = 10\ k\Omega$	>50,000 $R_L = 10\ k\Omega$	>50,000 $R_L = 10\ k\Omega$
Full temperature										
Input offset voltage (mV)	<6	<3	<5	<6	<3	<3	<7	<5	<3	<1
Input offset voltage (µV/°C)	NS[d]	<25	<20	NS	<15	<15	NS	<25	<15	<5
Input offset current (nA)	<500	<250	<500	<500	<20	<10	<5	<5	<0.4	<0.4
Input offset current (pA/°C)	NS	<500	<200	NS	<200	<150	NS	<40	<2.5	<2.5
Input bias current (nA)	<1,500	<600	<800	<1,500	<100	<75	<35	<25	<3	<3
Large-signal voltage gain[c]	25,000 to 70,000	25,000 to 70,000	50,000	>25,000	>25,000	>25,000	>25,000	>25,000	>25,000	>40,000
Output voltage swing[c] ($V_{p\text{-}p}$)	>20	>20	20	>20	>24 $R_L = 10\ k\Omega$	>24 $R_L = 10\ k\Omega$	>24	>26 $R_L = 10\ k\Omega$	>26 $R_L = 10\ k\Omega$	>26 $R_L = 10\ k\Omega$
Input voltage range (±V)	>8	>8	>10	>12	>15	>12	>12	>13.5	>13.5	>13.5
CMRR (dB)	>70	>80	>70	>70	>80	>80	>80	>80	>85	>96
Supply current (mA)	NS	<4.5	<4.5	<3.3	<2.5	<3.3	<2	<1	NS	NS
Unity-gain slew rate (V/µs)	NS	NS	0.5 typ	0.5 typ	0.5 typ	0.5 typ	2.5 typ	>0.1	0.2 typ	0.2 typ
Operating temp range (°C)	−55 to +125	−55 to +125	−55 to +125	−55 to +125	−55 to +125	−55 to +125	−55 to +125	−55 to +125	−55 to +125	−55 to +125
Operating supply voltage (±V)	9–15	9–15	9–15	15	20	15	15	20	20	20

[a] L. Altman, "Bridging the Analog and Digital Worlds with Linear ICs," *Electronics*, June 5, 1972 (Vol. 45, No. 12), Special Report (p. 85), copyright 1972, McGraw-Hill, Inc. All rights reserved.

[b] For description of device type, see Appendix III, where the extensive cross-reference listing identifies the type number by the italicized (or bold) part of the manufacturers' model number.

[c] $R_L = 2\ k\Omega$ unless otherwise specified.

[d] Not specified.

Table 4-1 is a quick-glance comparison chart for various typical OP-AMP specifications. Although it represents a fairly detailed comparison of widely used general-purpose OP AMPs, it is well—certainly at the outset—to concentrate on the few major characteristics previously mentioned in Section 4-5, after first identifying the popular type numbers that are listed. (Note that the manufacturer's identifying prefixes that were listed in Section 1-7 are omitted in the chart, since *practically all these general-purpose monolithic types* are *second-sourced* by a number of manufacturers, as can be seen by the emphasized numbers in the *cross-reference list* of OP AMPs given in Appendix III.)

Description of Listed OP-AMP Types

Type **709** (such as µA709), one of the early types, is suitable for operation over a fairly wide band of frequencies, but requiring, however, different values of two compensating capacitors (C_c) for various values of gain; thus, at closed-loop gain of 100 (40 dB), a bandwidth of more than 1 MHz is obtained with C_{c1} = 100 pF and C_{c2} = 3 pF (as recommended for that value of gain), resulting in a *gain–bandwidth product of over 100 MHz*.

Type **741** (*such as* µA741) *is internally compensated*, and is unconditionally stable at all values of gain without any external capacitor; the resulting f_c for 40-dB gain is 10 kHz, giving a *gain–bandwidth product for this condition of 1 MHz*.

Type **748** (*such as* µA748) *offers an extended bandwidth* over the 741 type, using only a single compensating capacitor (such as 30 pF) for all values of gain when a limited bandwidth (of around 10 kHz) is satisfactory. Otherwise, using a smaller C_c of 3 pF at the same gain of 40 dB, f_c is extended from 10 to around 100 kHz, with a resulting *gain–bandwidth product of 10 MHz*.

Type **101A** (such as LM101A) has the same simple (30-pF) single-capacitor compensation feature as the 748 type, but with *improved input specifications*.

Type **777** (such as µA777) and *type* **660** (such as SN52660) are termed *precision* OP AMPS (as are the others that follow, indicating usefulness in circuits requiring greater accuracy); they feature *smaller offset-error values*.

Type **770** (such as SN52770) and *type* **108A** (such as LM108A) are *super-beta transistor* types; they both feature *high input resistance*; additionally, the 770 type achieves a *higher slew rate* (also indicating better high-frequency response), while the 108A type exhibits still *further improvement in offset-error and drift characteristics*.

Other advanced types of OP AMPS that may be just mentioned at this point include the *FET-input types* (such as the µA740) and *programmable types* (such as the CA3080); these and other later types are discussed in Chapters 11 and 12 and listed in Appendix III.

4-7 INTERPRETING OPERATIONAL-AMPLIFIER DATA SHEETS

When selecting an OP AMP for a particular purpose, one is confronted with the difficulty of choosing from a large number of OP AMP types (even in this limited category of the wider field of linear ICs). For purposes of clarification, this chapter has concentrated on examples of typical specifications primarily for the widely used types (leaving highly specialized types for later); yet it is well to further subdivide these popular types according to three main kinds of applications (specific application details are covered in Chapter 5). Accordingly, in summarizing the importance of various specifications, examples of some suitable type numbers[2] are given next, along with those characteristics that are most pertinent for the *three general kinds of applications*, as follows:

A. For *voltage preamplification, within the audio-frequency band*, the main characteristics involve the following:
   1. Large-signal voltage gain (V/V).
   2. Input resistance (R_{in}).
   3. Output-voltage swing ($\pm V_{out}$).
   (*Examples:* types 741, 748, 101A, and 770.)
B. For *wider-band ac amplification*, additional consideration is given for the following characteristics:
   1. Open-loop frequency-response graph (for $G \times BW$ product).
   2. Frequency-compensation methods (C_c).
   (*Examples:* types 709, 748, 101A, and 777.)
C. For *precision dc amplification purposes* (as in instrumentation and analog computation) additional consideration is given for the following characteristics:
   1. Input-offset values of voltage and current (V_{io} and I_{io}).
   2. Input-offset drift with temperature (microvolts or picoamperes per degree Celsius).
   3. Common-mode rejection ratio (CMRR).
   4. Unity-gain slew rate (volts per microsecond).
   (*Examples:* types 660, 770, 777, and 108A.)

Keeping in mind the different requirements of these three kinds of amplification, that is, narrow-band, wide-band, and dc amplifiers, we can make a more sensible choice of a suitable OP AMP by giving proper weight to the various specifications. Accordingly, in presenting detailed application information for the schematic diagrams that follow in Chapter 5, a particular OP AMP type is

---

[2] See Appendix II for selection guide for OP AMPs, and also Appendix III for a cross-reference listing of device numbers from other manufacturers.

suggested with the understanding that other type numbers will also be suitable, as determined by the pertinent specifications.

4-8 APPROXIMATE VALUES OF TYPICAL OP-AMP SPECIFICATIONS

The numerical values listed for comparison in the chart of Table 4-1 can be very helpful in providing a preliminary estimate of "ball-park" figures for the various types of general-purpose OP AMPs. Referring back to the first two major characteristics (outlined previously in Section 4-5), the values of open-loop gain (A_{VOL}) and the gain–bandwidth product ($G \times BW$) are not given on the chart, since these figures are usually obtained from the graph of the open-loop frequency response (similar to Fig. 4.4 and generally found on the individual data sheet); however, an estimate of typical values for the $G \times BW$ product has been discussed in the previous section, where each type of OP AMP was described. (*Note:* The significance of the ratio of A_{VOL}/A_{VCL}, i.e., the *loop gain $A_o\beta$*, which is indicated on the plot of Fig. 4.4, is discussed later with regard to its effect on accuracy requirements.)

Likewise, the other two major characteristics [large-signal voltage gain (V/V) and input resistance (R_{in})] have also been estimated in the previous discussion of each type in Section 4-6. The characteristics that are discussed next have greater or smaller significance as specifications depending on particular applications.

Offset-Error Specifications

The specifications that are given for *input offset voltage V_{io}* (or offset current I_{io}) and the corresponding values of *offset drift* (microvolts or picoamperes per degree Celsius) are of negligible importance in the amplification of ac signals, but they do assume important significance in dc applications, *where precise dc information is required.* Such dc applications include dc instrumentation, analog computation, and, in some cases, digital interfacing.

For these dc applications, the values of offset error given in the chart present a basis of cost–performance selection. The values shown for *input offset voltage (V_{io} or sometimes V_{ios})* range from around 5 mV for the 741 type down to less than $\frac{1}{2}$ mV for the 108A type. However, a provision for the control of *offset null* is generally shown in the data sheet, so that this error can initially be adjusted to zero; consequently, greater importance is attached to the *offset-voltage drift specification.* Values for this drift characteristic are seen to range from 25 down to 5 $\mu V/°C$.

In a similar manner, the specifications for *input offset current (I_{io})* and corresponding *drift* (in picoamperes per degree Celsius) assume greater significance when working with a high-impedance source, where it is desirable to keep the resulting voltage drop to a minimum.

Full-Temperature Operation

The values shown in the lower portion of the chart apply for all the OP-AMP types over the operating-temperature range of $-55°$ to $+125°$C (the *military grade*), as opposed to the room-temperature ($25°$C) values shown in the first portion of the chart. A lower-grade (and lower-cost) type of $0°$ to $70°$C is generally available for these popular OP-AMP types, and this is indicated by various suffixes to the type number (such as C for the *commercial* grade) or by other devices. In many cases, an *industrial* grade is also offered as an in-between grade; for example, in the case of National types, we have LM101A (military), LM201A (industrial), and 301A (commercial) as various grades of the 101A type. It is well to emphasize here that the commercial grade ($0°$ to $70°$C) would likely be quite satisfactory for the great bulk of general-purpose OP-AMP applications, as can be determined from the manufacturer's data sheet for that type.

Output Specifications

The amount of voltage output that may be expected (over full-temperature operation) is shown as *output-voltage swing*, and is seen in all cases to exceed 20 V peak to peak (±10V), under the condition where the load resistance (R_L) is greater than 2 kΩ in most cases (but greater than 10 kΩ in the precision types). Since the *large-signal voltage gain* also exceeds 25,000 in all these cases, the full output voltage can be obtained for input signals as small as about 1 mV peak-to-peak. The maximum for the *input-voltage* range is around ±10 V.

The preceding values indicate the great versatility of the OP AMP in its ability to provide almost any reasonable value of undistorted voltage amplification within its *power-dissipation limits*. As a rule of thumb, the value of 500 mW for the 748 type can be taken as representative of the maximum power-dissipation value for most general-purpose OP AMPs. This indicates that this type of general-purpose OP AMP must be regarded primarily as a *voltage preamplifier*, since the amount of current in the load can be only a few milliamperes at best, in order to avoid distortion from saturation effects. In cases where greater amounts of current are required, an additional high-current (or booster) type of OP AMP (or in extreme cases, an external power transistor) is usually added in cascade with the general-purpose OP AMP, as is discussed later under the topic of power amplifiers. In the opposite situation, where the requirement for output current is not important, there are *micropower* OP AMPs, which draw very little power from the supply. These form a special type, as discussed later, for providing usable amplification from much smaller power sources, much less than the usual ±15 V and the 1 to 2 mA that are typically used in the general-purpose types.

Common-Mode Rejection Ratio

This characteristic measures the ability of the OP AMP to reject interfering signals (such as hum pickup) that are equally present at both inputs. With this ratio exceeding 70 dB in all cases, the gain for the desired signal is assured to be at least 70 dB or about $3,000^3$ times greater than the gain for the unwanted common-mode signal.

Unity-Gain Slew Rate

Measuring the ability of the OP AMP in following fast-changing signals, this characteristic assumes importance in fast-rise pulse circuits and other such signals emphasizing high-speed operation (such as in many digital-interface types). This specification is discussed later at greater length for cases requiring values greater than the 100- to 500-mV/μs rates listed in Table 4-1. In Appendix III these types are designated as "Fast" or "High Slew Rate" types.

Single and Dual Power Supplies

The usual power source for most OP AMPs is a dual supply (generally ±15 V, or, for operation by convenient transistor batteries, ±9 V). The connection for ground is usually taken from the mid-point connection of $V+$ and $V-$; this connection provides the ground point for single-ended inputs and also for single-ended outputs (even though this ground point is often assumed and not shown on the diagram). In most commercial cases, the dual supply is regulated, often very simply by use of a three-terminal regulator, as described in Chapter 9.

The use of this *dual (or "split") power supply* provides a number of important advantages in the circuit design of an OP AMP. Among other things, it simplifies the *internal biasing arrangement.* In addition, it satisfies an important property needed for dc amplification: it ensures that there is essentially *zero output for zero input* (disregarding the inherent offset voltage, which can easily be eliminated by an offset-null provision). These advantages of the dual power-supply systems generally outweigh the disadvantages of requiring an extra supply, with its attendant additional cost, especially since it enables the OP AMP to be used in a versatile fashion for both dc and ac amplification.

A special modification of the conventional dual-supply OP AMP, which is specifically designed for single-supply operation, is available for cases where

---

[3] 3,160 times from the decibel conversion chart in Appendix I.

only ac amplification is involved. This type, the current-difference OP AMP, used extensively in quad OP AMPs with single supply, is discussed in Section 5-16.

QUESTIONS

4-1. The effect of using a larger feedback resistor (R_f) causes the no-signal output voltage of the OP AMP to
(a) Increase
(b) Decrease
(c) Remain the same

4-2. Using a smaller input resistor (R_i), with other factors remaining the same, will generally cause the no-signal output to
(a) Increase
(b) Decrease
(c) Remain the same

4-3. The main advantage in using the noninverting terminal for the input of an OP AMP (disregarding the polarity of the output) is
(a) Greater gain
(b) Greater input impedance
(c) Greater output

4-4. The offset null terminals of an OP AMP are primarily used to reduce the offset voltage in
(a) dc applications
(b) ac applications
(c) High-frequency applications

4-5. An outstanding advantage for simplicity of operation of an OP AMP is offered by the internal compensation feature in
(a) Type 709A
(b) Type 741
(c) Type 748

4-6. The effect of a smaller input bias current (I_B) causes the input offset voltage (V_{io}) to
(a) Increase
(b) Decrease
(c) Remain the same

4-7. A type 741 OP AMP has an input offset voltage $V_{io} = 5$ mV, and is used with an input resistor R_1 of 10 kΩ, and feedback resistor $R_f = 500$ kΩ. The quiescent output voltage (V_o) with zero input signal is
(a) 5 mV

(b) 50 mV
(c) 250 mV
(d) 500 mV

PROBLEMS

4-1. In the circuit for measuring the input offset voltage (V_{io}) of an OP
AMP (Fig. 4.5), find the value of V_{io}
(a) For type 741, when the output voltage $V_o = 500$ mV
(b) For type 101A, when $V_o = 300$ mV
(c) For type 108A, when $V_o = 100$ mV
(d) If R_f is increased to 470 kΩ, what effect may be expected on
the corresponding output voltages?

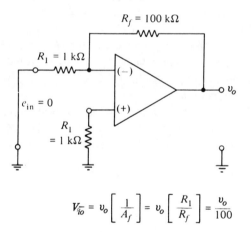

$$V_{io} = v_o \left[\frac{1}{A_f} \right] = v_o \left[\frac{R_1}{R_f} \right] = \frac{v_o}{100}$$

Figure 4.5 Measuring input offset voltage (V_{io}): with both inputs at ground,
the input offset voltage is V_o divided by the closed-loop gain.

4-2. In the basic OP-AMP circuit of Fig. 4.1, find the values asked for in
order to obtain a closed-loop amplification (A_{VCL}) of –100 in each
case.
(a) $R_1 = 2.2$ kΩ, $R_f =$
(b) $R_1 = 27$ kΩ, $R_f =$
(c) R_1 (min) $= 470$ Ω, $R_f =$
(d) R_f (max) $= 10$ MΩ, $R_1 =$

4-3. In Fig. 4.1, assume

$$A_{VOL} = -50,000$$

$$R_1 = 10 \text{ k}\Omega$$

Find the loop-gain (LG, as distinguished from the closed-loop gain A_{VCL}) for the following conditions:
(a) $R_f = 100 \text{ k}\Omega$, LG =
(b) $R_f = 2 \text{ M}\Omega$, LG =
(c) In which of the above cases is there a greater input impedance seen at the noninverting input?

4-4. Using the full-temperature specifications of Table 4-1, find the OP AMP
(a) With the smallest input offset voltage (V_{io}), and compare it with that of the type 741
(b) With the greatest common-mode rejection-ratio (CMRR), and compare it with that of the type 741
(c) For the same external circuit components in a given OP-AMP circuit, and using a load resistor (R_L) = 2 kΩ in each case, compare the gain obtained from type 101A with type 741.

4-5. Using the open-loop frequency-response curve of Fig. 4.4, find the approximate value of
(a) Closed-loop bandwidth (f_{CL}) at a gain of 10 times (20 dB)
(b) Closed-loop bandwidth (f_{CL}) at a gain of 1,000 times (60 dB)
(c) Compare the $G \times BW$ product of the two cases.

Chapter 5

General Operational Amplifier Applications

5-1 KINDS OF OPERATIONAL AMPLIFIER APPLICATIONS

The successful introduction and acceptance of the integrated-circuit version of the operational amplifier radically changed its role from its *original functions in analog computation.* In its discrete form, whether using tubes or transistors, the operational amplifier occupied the fairly lofty position of a high-performance (and correspondingly costly) dc amplifier, capable of performing the mathematical functions of summing, multiplying by a constant, and integrating analog signals in such a manner as to accomplish a solution of differential equations. It thus served as a powerful mathematical tool in working with the highly flexible electrical analog forms that were the simulated equivalents of bulky or otherwise awkward physical processes.

With the breakthrough in techniques for the fabrication and circuit design of integrated circuits, the operational amplifier became available as a more reliable and considerably less costly device; since then it has progressed rapidly into becoming the "versatile OP AMP,"[1] closely approaching the ideal of a "universal amplifier," with its many uses as a building-block device.

In its present IC form, there are obvious physical constraints that limit the IC OP AMP with respect to such things as *high power* and *LC circuits*, conditions

[1] M. Kahn, *The Versatile Op Amp*, Holt, Rinehart and Winston, Inc., New York, 1970.

that generally call for external (or hybrid) additions to the circuit. Yet, even in these cases, developments in design strategy are gradually encroaching on such limitations. For example, advances have been made toward higher output-current amplifiers and also in gyrator circuits that have been developed to simulate inductors.

The applications given in this chapter are necessarily restricted to typical examples. They illustrate some of the wide diversity of basic OP AMP uses in various *amplification and oscillating functions using general-purpose OP AMPS*. A separate chapter discusses instrumentation OP AMPs, followed by chapters discussing many other linear IC functions, such as comparators, regulators, tuned circuits, active filters, and interfacing with digital circuits.

5-2 CHOICE OF OPERATIONAL AMPLIFIER IN TYPICAL APPLICATIONS

Where a particular type of OP AMP is suggested in the application schematics, it should be noted that a number of suitable substitutions are possible. The ultimate choice will obviously also include the relative-cost factor balanced against the specifications, for example, Table 4-1, and the relative weights attached to them, as previously discussed in the sections that follow the table. (Possible equivalents other than those in the table may be found in the cross-referenced list in Appendix III and the selection guide in Appendix II.)

For constructing prototypes of the applications, practical details are given in Chapter 6 for *testing and breadboarding* the application examples.

5-3 BASIC CIRCUIT CONFIGURATIONS

In the application information that follows, we start with the two configurations for *voltage amplification* (inverting and noninverting) for which the fundamental gain relation (R_f/R_1) was discussed in Section 4-2. In addition, arrangements for analog computation include configurations for *summing, multiplying by a constant, integrating*, and *differentiating*. Also, *oscillator functions* are realized by means of a configuration to provide positive feedback. Other *nonlinear functions* (such as *detectors*, but still done with the linear ICs) are also given. These circuits, together with their modifications, result in the typical applications that follow.

5-4 INVERTING VOLTAGE AMPLIFIER

The voltage amplifier of Fig. 5.1 is a fairly simple *general-purpose preamplifier*. The type 741 OP AMP is internally compensated and, as such, is the easiest to

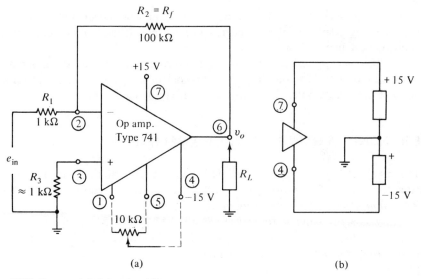

(a)

(b)

NOTE: Pin nos. apply to 8-pin package (either
8-pin metal can or 8-pin mini-DIP types).

Figure 5.1 Inverting amplifier: (a) circuit for general-purpose preamplification
(voltage gain of 100) using type 741 (internally compensated) OP
AMP, with offset null adjustment shown by dashed lines; (b) arrangement for dual power supplies. (*Fairchild Camera and Instrument Corporation*)

use for applications *within the audio-frequency range.* We can obtain a voltage
gain of any reasonable value (as low as unity gain and up to a large-signal gain of
50,000) in a very straightforward manner by the ratio of feedback resistor R_f
to input resistor R_1; in the example, R_f/R_1 is 100:1 for a gain of -100. The
other resistor, R_3, at the noninverting terminal 3, should theoretically be equal
to the parallel combination of R_f and R_1 $[R_1 R_f/(R_1 + R_f)]$; in this case, so
close (within 1 percent) to R_1 alone, we can use the same value at both input
terminals and still retain minimum offset

Input resistance for the inverting connection is fairly low, generally taken
as equal to input resistor $R_1 = 1,000$ Ω in this case (the next section, on the
noninverting connection, yields a much higher input resistance).

Frequency response, obtained from the open-loop frequency-response
curve, (see Fig. 4.4), will show a bandwidth or f_c(-3 dB) of approximately 10
kHz for this gain of 100 (or 40 dB). At a lower gain of 10, the bandwidth increases correspondingly, to about 100 kHz, as seen in Fig. 4.4.

Power-supply connections [part (b) of the figure] show the dual-supply
that is generally used with OP AMPs. Note that the IC itself does not have a
terminal for the common ground connection; it is made externally at the junction of the positive and negative supplies. These (V+ and V- values) may be

typically anything between ±9 and ±15 V, depending on the amount of output-voltage swing desired. (Values smaller than ±9 V may be used, if desired, with a consequent reduction in gain.)

Offset-null provision is provided (as shown by dashed lines in the figure) by the use of a 10-kΩ potentiometer. This control is adjusted, when needed for dc amplification, to obtain zero output for zero (grounded) input.

5-5 NONINVERTING VOLTAGE AMPLIFIER

The circuit of Fig. 5.2 for a noninverted output is quite similar to the inverting circuit of Section 5-4, providing much the same gain but a considerably *higher input impedance* (in the high-megohm range).

By analysis (similar to that in Section 4-2)[2] the gain is found to be $A_{VCL} = 1 + (R_f/R_1)$, and the closed-loop input impedance $= Z_i(1 + LG)$ or approximately the original open-loop input impedance multiplied by the loop gain (LG).

The use of the type 748 OP AMP in this circuit requires an external frequency-compensation capacitor (C_c), but because it has external compensation, it allows an *extended bandwidth* (beyond the audio range to around 100 kHz), which is obtained quite simply by reducing the nominal 30 pF capacitor (C_c) to 3 pF in this case. (In-between values can be obtained from the frequency-response curves given in the 748 data sheet.)

The *offset-null circuit* (shown in dashed lines for dc applications) uses a 5-MΩ potentiometer in this case, but an alternative circuit for a 25-kΩ potentiometer is shown in part (b) of the figure.

The *power supply* again shows a ±15-V dual supply, which can be lower (±9 V, or even less with reduced gain). Where only a single supply is available, an alternative circuit is shown in part (c) of the figure for a 20-V supply, center tapped at +10 V by two equal bleeder resistors (R).

Either of these widely used OP AMPs (types 741 and 748) is generally useful for almost any reasonable value of gain, especially for ordinary *ac pre-amplification*, where offset-error drifts with temperature are not particularly significant.

Modified Versions of Voltage Amplifier

Where there are stricter requirements for the amplifier, other refined OP-AMP types with tighter specifications are suggested in some typical examples. Like-wise, there are instances where additional controls are desired (such as for vary-

---

[2] For derivation, see G. J. Deboo and C. N. Burrous, *Integrated Circuits and Semiconductor Devices*, 2nd ed., McGraw-Hill Book Company, New York, 1977.

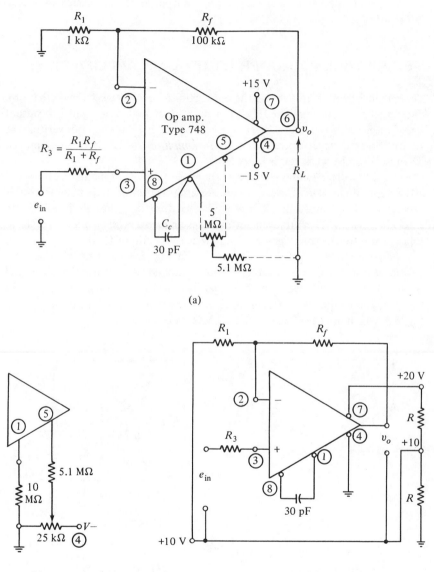

Figure 5.2 Noninverting amplifier: (a) voltage gain of 100 circuit at high input impedance and bandwidth (≈10 kHz) can be extended by reducing C_c (e.g., when $C_c = 3$ pF, $BW ≈ 100$ kHz); offset null needed for dc amplification shown in dashed lines; (b) alternate offset null arrangement; (c) circuit for single power supply. (*Fairchild Camera and Instrument Corporation*)

ing gain, controlling tone, increasing current output, or the like); such additional information is summarized in Chapter 11.

5-6 VOLTAGE FOLLOWER (UNITY-GAIN AMPLIFIER)

The circuit of Fig. 5.3(a) shows the connections for a *voltage follower* that produces unity gain (noninverted), with a high input impedance and low output impedance. Using the internally compensated type 741 OP AMP, this very simple circuit serves quite effectively as a *buffer amplifier*, which is often employed when isolation between stages is needed.

A *fast voltage-follower* circuit is shown in part (b) of the figure, using the 101A type of OP AMP. The usual compensation capacitor (C_{c1} = 30 pF) of this type is used in conjunction with a second capacitor (C_{c2} = 300 pF, in series with a 10-kΩ resistor). This arrangement provides an *increased slew rate* of 1 V/μs to follow abruptly changing signals and a power bandwidth of 15 kHz.

Note that both cases of the circuits in Fig. 5.3 show no provision for balancing out any dc offset errors. If this is desired, we can use the offset-null provisions previously given—for the type 741, using the 10-kΩ pot shown in Fig. 5.1, and for the type 101, or using the same provision as for the 748 type in Fig. 5.2, which used a 5-MΩ pot and a 5-MΩ series resistor.

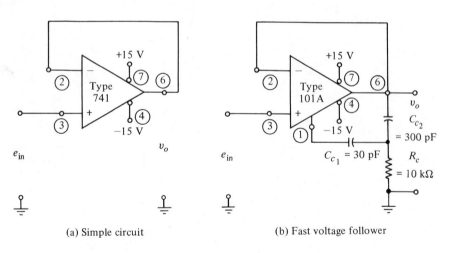

(a) Simple circuit   (b) Fast voltage follower

NOTE: Pin nos. apply to 8-pin package (either
8-pin metal can or 8-pin mini-DIP types).

Figure 5.3 Voltage-follower (unity-gain) circuit: (a) simple circuit suitable for most uses; (b) connections for fast voltage-follower. [(a) *Fairchild Camera and Instrument Corporation* and (b) *National Semiconductor*]

5-7 SUMMING AMPLIFIER

The OP AMP is widely used for combining a number of inputs in an additive manner as, for example, in the analog computer. Using an inverting mode, the summing amplifier of Fig. 5.4 will produce an *output proportional to the negative of the sum of the inputs:*

$$v_o = -(k_1 e_1 + k_2 e_2 + k_3 e_3).$$

Inserting the value for each constant k, the equation becomes

$$v_o = -\left(\frac{R_f}{R_1} e_1 + \frac{R_f}{R_2} e_2 + \frac{R_f}{R_3} e_3\right),$$

or, from Fig. 5.4,

$$v_o = -(e_1 + 4e_2 + 2e_3).$$

Since the junction point of the input resistors and the negative feedback resistor is forced to practically zero potential (virtual ground), the various inputs

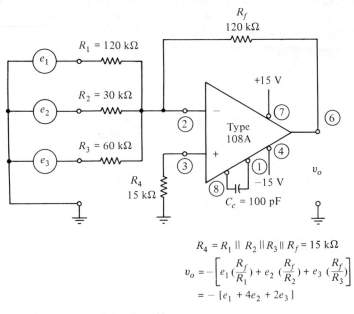

$$R_4 = R_1 \parallel R_2 \parallel R_3 \parallel R_f = 15 \text{ k}\Omega$$

$$v_o = -\left[e_1\left(\frac{R_f}{R_1}\right) + e_2\left(\frac{R_f}{R_2}\right) + e_3\left(\frac{R_f}{R_3}\right)\right]$$

$$= -[e_1 + 4e_2 + 2e_3]$$

NOTE: Pin nos. apply to 8-pin package (either
8-pin metal can or 8-pin mini-DIP types).

Figure 5.4 Summing amplifier: the output (v_o) is proportional to the negative of the sum of the inputs.

are effectively isolated from each other. In the circuit shown, a type 108A OP AMP is used by virtue of its combination of high-accuracy and high-input-impedance properties; when desired, however, a general-purpose OP AMP (such as type 741) can also be used at a small sacrifice in the accuracy of the summation.

Subtracting Function

The summing amplifier can add both positive and negative voltages, thus forming the equivalent of a *subtracting function* for general-purpose use. However, when voltages of the same relative polarity are to be subtracted, the OP AMP is generally used in the differential (or *dual-input*) mode, where input signals are applied to both inverting and noninverting terminals. This latter mode provides an output proportional to the difference of the inputs, as described in Section 5-15 and under differential OP-AMP applications in Instrumentation Amplifiers in Chapter 11.

5-8 INTEGRATING AMPLIFIER

The amplifier of Fig. 5.5 accomplishes the integration of an input signal by employing a capacitor (C_f) as its primary feedback impedance element (Z_f). Disregarding the shunting resistor (R_2) across it, for the time being, we may apply the basic OP-AMP relation in the *p-operator* form[3] where

$$p = \frac{d}{dt}, \qquad \frac{1}{p} = \int dt,$$

and
$$Z_1(p) = R_1, \qquad Z_f(p) = \frac{1}{pC_p}.$$

Then, substituting the p-operator impedances,

$$v_o = -\frac{Z_f(p)}{Z_1(p)} e_1 = -\frac{1}{pR_1C_p} e_1$$

and, since $1/p$ signifies integration,

$$v_o = -\frac{1}{R_1C_f} \int e_1 \, dt,$$

[3] In the complex s plane, the symbol s is used for p, giving the more recognizable expression for $v_o(s) = 1/RCs$.

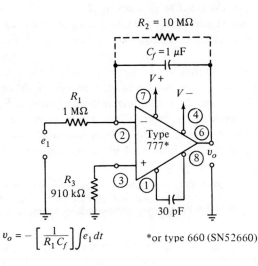

$$v_o = - \left[\frac{1}{R_1 C_f} \right] \int e_1 \, dt$$

*or type 660 (SN52660)

(a)

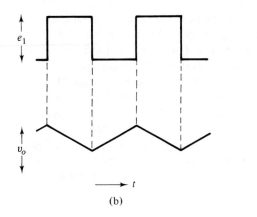

(b)

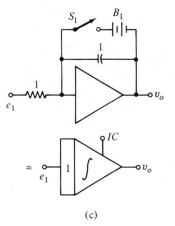

(c)

NOTE: Pin nos. apply to 8-pin package (either
8-pin metal can or 8-pin mini-DIP types).

Figure 5.5 Integrating amplifier: (a) output v_o is the negative of the time
integral of the input e_1 times a constant, usually made unity ($R_1 =$
1 MΩ, C_f = 1 μF); (b) triangular wave is the integral of a square-wave
input; (c) initial conditions are shown, either by switch S_1 (normally
closed) and voltage B_1, or in shorthand form below, by terminal IC
(for initial condition). (*Fairchild Camera and Instrument Corporation*)

or the output will be the *negative of a constant times the time integral of the input*. Thus, as shown in part (b) of the figure, a square-wave input will be integrated to yield an inverted triangular-wave output with a time constant of $R_1 C_f$ seconds.

The resistor R_2, shunting C_f, is shown in dashed lines, to be used when needed to prevent saturation of the amplifier, especially if the OP AMP has substantial voltage offset. Since the integrating amplifier (without R_2) is operating open loop for direct current, the input offset voltage would be integrated, and the output would tend to rise in a ramp fashion to possibly saturate the amplifier (in either direction). This is avoided when the dc gain is limited by the shunt resistor R_2.

Because of the multiplying effect on the input offset voltage, a *precision type of OP AMP* is indicated for accuracy in integration. Consequently, the example for this application shows a type 777 OP AMP (or a similar precision type).

The integrator finds wide application in the analog computer for solving differential equations in simulation applications. For example, in equations for accelerated motion, the expression $d^2 y/dt^2$ is easily solved for the displacement y by a double integration,[4] yielding $-dy/dt$ after the first integration and $+y$ after the second integration. In analog computation (and where saturation effects are avoided by large output-voltage swings without the use of R_2), the values of R_1 and C_f are usually taken as 1 MΩ and 1 μF, respectively, making the constant term $1/R_1 C_f$ equal to unity; this simplifies the computation. Also, when input current offsets are small enough to be neglected, it is usual to eliminate resistor R_3 at the noninverting terminal (where it usually has the value of the parallel resistance of R_1 and R_f in order to balance the input currents, as previously discussed). With R_3 eliminated, the symbolic representation of the integrator takes either of the two simple forms shown in part (c) of the figure. The provision for *setting initial conditions* at the appropriate integrator is shown by the normally closed switch S_1, which applies the required voltage from battery B_1 and which is opened at the start of the integration.

5-9 DIFFERENTIATING AMPLIFIER

The *differentiation function* is accomplished by interchanging the input resistor and feedback capacitor of the OP AMP, as shown in Fig. 5.6. Examining the *p*-operator form of the basic OP-AMP relation, as in Section 5-8 (and neglecting

---

[4] For a worked-out example, see S. Prensky, *Electronic Instrumentation*, 2nd ed., Prentice-Hall, Inc., Englewood Cliffs, N.J., 1971.

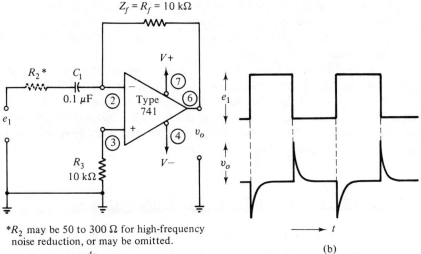

*R_2 may be 50 to 300 Ω for high-frequency noise reduction, or may be omitted.

$$v_o = -\left[R_f C_1 \right] \frac{de_1}{dt}$$

(a)

NOTE: Pin nos. apply to 8-pin package (either 8-pin metal can or 8-pin mini-DIP types).

Figure 5.6 Differentiator amplifier: (a) the output v_o is the negative of the time derivative of the input e_1 times a constant $(R_f C_1)$, which is 1 millisecond (1×10^{-3}) in this example (see text for optional R_2); (b) peaked output waveform resulting from differentiation of square wave.

the dashed-line connection), we find

$$v_o = -\frac{Z_f(p)}{Z_1(p)} e_1 = -\frac{R_f}{1/C_1 p} e_1 = -(R_f C_1 p) e_1$$

$$= -R_f C_1 \frac{de_1}{dt}$$

For the ideal amplifier, this relation yields the *negative of the differential of the input voltage times a constant*. It should be noted, however, that the capacitor impedance is now in the denominator of the formula, indicating that the differentiator will give an increased output at higher frequencies. Thus the *high-frequency noise problem* of the differentiator becomes *worse with increased frequency*, opposite to the integrator. For this reason, the analog computer seldom uses the differentiator, preferring to arrange the differential equations so that they require integration for their solution.

(Be careful here not to confuse the *differentiator*, discussed in this section, with the *differential amplifier*; the first performs a differentiating function to produce the *mathematical derivative* of the input, whereas the differential amplifier, which is discussed later under instrument amplifiers, is an amplifier with a double-ended input which produces an output that is proportional to the *difference between the two input signals.*)

The differentiating amplifier does, however, find use in *wave-peaking* and other *wave-shaping* circuits; in such cases the use of the series resistor R_2 (shown in dashed lines in the figure) is justified to reduce high-frequency noise at the expense of attenuation. As an example of the use of the differentiation for wave shaping in pulse circuits, part (b) of the figure shows how a square-wave input produces a sharply peaked wave at the output.

5-10 SQUARE-WAVE GENERATOR (MULTIVIBRATOR)

There are many IC circuits capable of providing free-running (astable) multivibrator action, producing an output of rectangular or square waveforms. These *relaxation oscillators* are essentially amplifiers that oscillate by means of *positive feedback, with the frequency controlled by a given RC time constant.* (Generation of a sine-wave output is covered in the next section.)

The effect of the positive feedback in a multivibrator circuit is to produce regenerative action that alternately affects each of two active devices, thus initiating abrupt transitions between on and off states. This results in the fast-rising and fast-falling waveform of a rectangular wave, having a corresponding *period* of T_1 (on state) plus T_2 (off state). This type of rectangular waveform is usually called a square wave, even though asymmetric (the symmetrical square wave must have $T_1 = T_2$, and this symmetry can usually be obtained by simple adjustment of resistor values).

Although the recurrent switching (or astable) action of the multivibrator can be accomplished with digital IC gates by cross coupling, the OP-AMP circuits shown in Fig. 5.7 are preferred for providing a significantly greater output swing (easily up to 20 V peak to peak).

A simple circuit using a minimum of external components is shown in Fig. 5.7(a), employing a general-purpose (type 101A) OP AMP. This circuit is reliably self-starting, with the frequency adjustable by the value of C_1 within fairly wide limits.

A more complex circuit, shown in part (b) of the figure, has additional features in extending the low-frequency limits and additionally providing for a choice of either a low-impedance output or an output clamped to ±6.9 V by the back-to-back connection of the zener diodes. Here, again, stable operation at various frequencies can be obtained by adjustment of the C_1 capacitor.

The essential property of square waves lies in the fast rise times (and fall times) of the transitions. These abrupt step-voltage signals are the starting point for various other kinds of waveform generation, including such waveforms as

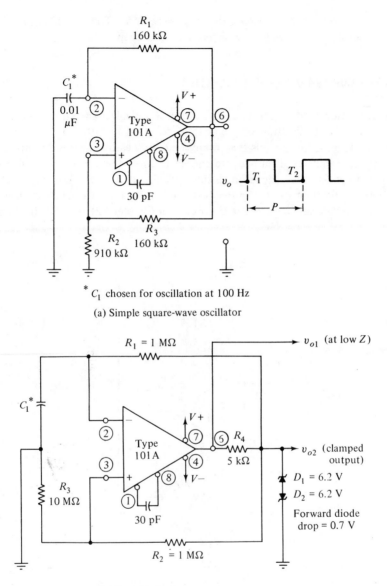

* C_1 chosen for oscillation at 100 Hz

(a) Simple square-wave oscillator

* Adjust C_1 for frequency

(b) Low-frequency square-wave oscillator

NOTE: Pin nos. apply to 8-pin package (either
8-pin metal can or 8-pin mini-DIP types).

Figure 5.7 Square-wave (multivibrator) generator: (a) simple circuit using
minimum components; (b) low-frequency version allows choice of
low-impedance or clamped output. (*National Semiconductor*)

triangular, *ramp*, and assorted *pulse-type waveforms*. The use of linear ICs in assembling such *function generators* is covered in Chapter 12.

5-11 SINE-WAVE OSCILLATORS

The generation of the basic sinusoidal waveform involves an extra consideration for amplitude stabilization of the output, to avoid distortion in the purity of the sine wave. Although this is accomplished fairly easily in a low-amplitude *LC* circuit, the examples given here are chosen to avoid the use of awkward inductors, and concentrate instead on flexible *RC* circuits that are more practical for OP-AMP applications (an exception is the use of dual OP AMPs in a *gyrator circuit* to simulate an inductor; it is covered in Chapter 12). The *RC* frequency-

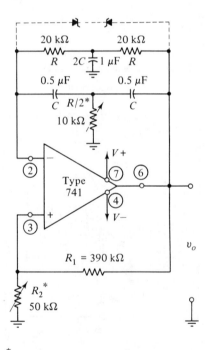

*Controls for trade-off between amplitude-stabilization and reduced distortion

– – – Optional amplitude-limitation

$$f_o = \frac{1}{2\pi RC}$$

(a) Twin- T Circuit

Figure 5.8   Sine-wave oscillators: (a) twin-T type of oscillator; (b) Wien-bridge type of oscillator; (c) Wien-bridge circuit of (b) redrawn to show bridge arrangement (see footnote references in text).

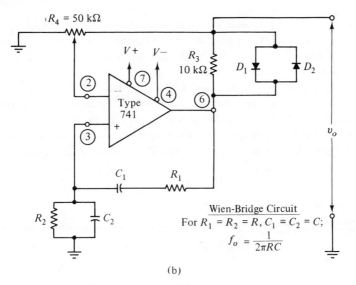

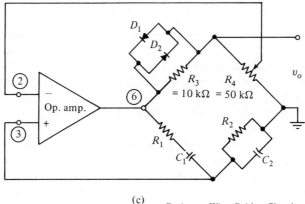

(b)

Wien-Bridge Circuit

For $R_1 = R_2 = R$, $C_1 = C_2 = C$;

$$f_o = \frac{1}{2\pi RC}$$

(c)

NOTE: Pin nos. apply to 8-pin package (either 8-pin metal can or 8-pin mini-DIP types).

Redrawn Wien-Bridge Circuit of (b)

Figure 5.8 (*Contd.*)

determining elements employed in the examples are of two basic types, the *twin-T type* first, followed by the *Wien-bridge* circuit.

Twin-T Type

In the circuit of Fig. 5.8(a),[5] two T-sections in parallel (twin-T) are connected in the negative feedback loop. The oscillation frequency is based on the twin-T

[5] J. D. Lenk, *Handbook of Simplified Solid-State Circuit Design*, Prentice-Hall, Inc., Englewood Cliffs, N.J., 1971.

property of zero transmission at the parallel-resonance frequency, $1/(2\pi RC)$. Since positive feedback is supplied by the voltage-divider action of R_1 and R_2, oscillation takes place at this "pole" frequency, where the negative feedback through the twin-T is at its minimum.

For the values shown for R and C [RC product = $10(10^{-3})$ or 0.01], the oscillation frequency is calculated as 16 Hz. Any other combination giving this RC product can also be used for this frequency (with the provision that R should not exceed 2 MΩ). Likewise, the values for other frequencies are easily obtained from the *inverse relation of the RC product to the oscillation frequency*, as shown in Table 5-1.

TABLE 5-1

Component Values for $f_o = \dfrac{1}{2\pi RC}$

$f_o = \dfrac{1}{2\pi RC}$	R (kΩ)	C (μF)	RC Product
16 Hz	20	0.5	$10(10^{-3})$
320 Hz = (16 × 20)	10	0.05	$0.5(10^{-3}) = 10(10^{-3})/20$
1.6 kHz = (16 × 100)	2	0.05	$0.1(10^{-3}) = 10(10^{-3})/100$

To assure a stable circuit that is self-starting, the value of R_2, controlling the amount of feedback, is made variable by a 50-kΩ control (nominal value is $2R$, or 40 kΩ) in conjunction with the fixed resistor R_1 having a value around $10R_2$ (here the closest standard value to 400 kΩ, or 390 kΩ).

Amplitude limiting, when necessary, is accomplished by the back-to-back connection of zener diodes, shown in dashed lines. By making use of the nonlinear resistance of the diodes, the tendency for the amplitude to increase is opposed by the effect of decreased diode resistance, as it approaches the knee of its curve. A rule-of-thumb value for the zener voltage is about 1.5 times the undistorted peak-to-peak value of the output sine wave.

Wien-bridge Circuit[6]

The Wien-bridge circuit enjoys popular use in commercial signal generators, where advantage is taken of the fact that it is conveniently tunable over a wide range of frequencies.[7] In the circuit of Fig. 5.8(b), the bridge arrangement is seen to consist of a series connection of $R_1 C_1$ followed by a parallel combination of $R_2 C_2$. With $R_1 = R_2 = R$ and $C_1 = C_2 = C$, the path has the property of

---

[6] G. E. Tobey, L. P. Huelsman, and G. G. Graeme, *Operational Amplifiers: Design and Applications*, McGraw-Hill Book Company, New York, 1971.

[7] Details of operation of the basic Wien bridge are given in Prensky, *Electronic Instrumentation*.

zero phase shift at a single frequency, $f_o = 1/(2\pi RC)$. Since this combination is used as the positive feedback element, oscillation occurs at this frequency of zero phase shift.

In its original form, amplitude stabilization in the Wien bridge was traditionally done in the path of additional negative feedback by means of the nonlinear resistance of a tungsten lamp. In this circuit, however, this function is obtained from back-to-back diodes, which are arranged to shunt the 10-kΩ resistor (R_3), which is in the negative feedback path, in series with the 50-kΩ distortion control (R_4). Here, again, as the output amplitude tends to increase, the beginning conduction of the diodes lowers their impedance, thus opposing the increase by producing more negative feedback.

The values of R and C for obtaining the desired frequency, which were given in Table 5-1, also apply here, since the same formula applies, $f_o = 1/(2\pi RC)$.

In part (c) of the figure the Wien-bridge circuit of part (b) is redrawn in a form to show the bridge structure more clearly.

5-12 LINEAR RECTIFIERS

For the rectifying function of changing ac signals to undirectional (pulsating dc) signals, the OP AMP offers a simple method, shown in Fig. 5.9, for overcoming the troublesome nonlinear properties of the rectifying diode used in this process. In the ordinary diode rectifier (whether half or full wave), diode conduction does not start until the ac voltage has risen to the breakover point, around 0.6 V (for the usual silicon diodes), and then provides current that has a nonlinear (exponential) relation to the input voltage.

Both of these difficulties are overcome by the use of the OP AMP in the linear half-wave rectifier of Fig. 5.9(a). Here, for small instantaneous negative voltages (less than 0.6 V), the amplifier is operating practically open loop, and so a relatively large positive voltage appears at the output end of the feedback elements, forcing the diode into forward conduction. In a similar manner, for various ac inputs the gain of the OP AMP automatically adjusts itself to keep the summing voltage at virtual ground. Thus the instantaneous values of v_o and e_1 are kept practically equal to each other, assuring a good linear relation. For this reason, the rectifier circuit of the OP AMP as shown is called a *precision* or *linear half-wave rectifier*.

A similar scheme may be used to provide *full-wave rectification*, as in an ac voltmeter, as shown in Fig. 5.9(b). The meter sensitivity primarily determines the ohms-per-volt sensitivity of the resulting ac scale, with R_{cal} made variable to help in the calibration. Thus, roughly speaking, a 0- to 1-mA dc meter can provide a sensitivity around 1,000 Ω/V; progressively greater sensitivities can be obtained by using smaller full-scale dc values—coming down to a 0- to 50-μA meter for a sensitivity of 20,000 Ω/V.

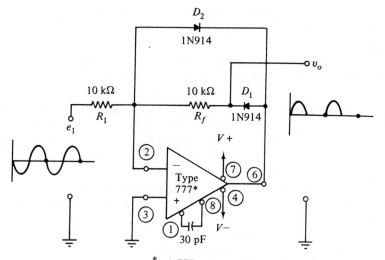

* μA 777, or type 660 (SN52660) (see Table 4-1)

(a) Linear half-wave rectifier

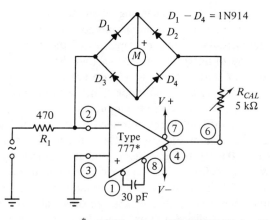

* μA 777, or type 660 (SN52660) (see Table 4-1)

(b) Full-wave voltmeter rectifier

NOTE: Pin nos. apply to 8-pin package (either
8-pin metal can or 8-pin mini-DIP types).

Figure 5.9 Linear rectifiers: (a) half-wave rectifier for more linear rectification, using precision type of OP AMP; (b) full-wave rectifier for more linear scale of ac voltmeter.

A higher input impedance can be obtained from more complex circuitry or, as an alternative, by using another OP AMP to serve as a unity-gain buffer before the rectifying OP AMP. In any case, both the linearity and the sensitivity of ac rectification are improved by the amplifying ability of the OP AMP. This is an obvious advantage over nonelectronic voltmeters where, for example, a multimeter having $20,000\text{-}\Omega/\text{V}$ value for direct current comes out with a much poorer figure of around $5,000 \ \Omega/\text{V}$ for ac measurements, thus necessitating a special *nonlinear* scale for the low ac ranges.

5-13 PEAK DETECTOR

It is often necessary to measure the ac voltage of signals having nonsinusoidal waveforms, and in such cases we cannot rely on the root-mean-square reading shown on the ordinary ac voltmeter, since it is generally calibrated on the basis of a pure sine-wave input. Such instances occur frequently with pulse measurements, with square and sawtooth waves, or even when dealing with greatly distorted sine waves. In such cases we are more interested in *peak measurements* (either single peak or peak-to-peak values).

An OP-AMP equivalent of a discrete peak detector is shown in Fig. 5.10. In part (a), the discrete circuit responds to the voltage developed by the fast charging of capacitor C by the positive peak, while very little voltage is lost by the

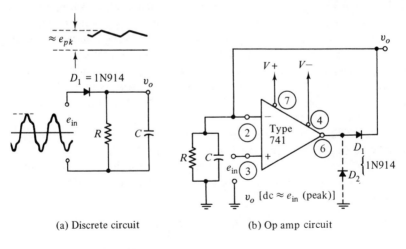

(a) Discrete circuit   (b) Op amp circuit

Pin nos. apply to 8-pin package (either
ietal can or 8-pin mini-DIP types).

Figure 5.10 Peak detector: (a) discrete circuit; (b) OP AMP circuit, producing a dc output voltage v_o, reasonably close to the positive peak of the signal.

relatively slow discharge of the capacitor through load resistor R when the diode is reverse biased on the negative swing, as shown by the output waveform.

The OP-AMP version of the peak detector, in part (b) of the figure, allows considerable improvement. Although a type 741 general-purpose OP AMP is shown, critical applications will profit from the use of an instrumentation type of OP AMP (discussed in Section 5-15) with advantages of an FET input and/or high slew rate. For ordinary applications, the output v_o is a dc voltage reasonably close to the amplitude of the positive peak of the input. Specific values for R and C depend upon keeping the RC time constant within a reasonable relation to the period of the input signal. (Where peak-to-peak voltage outputs are desired, the equivalent of a voltage-doubler circuit is used.)

The optional use of a second diode D_2, as shown by dashed lines in the figure, is added where it is necessary to avoid saturation of the OP AMP on the negative excursions of a large input signal. Also, when it is necessary to avoid loading the capacitor, a voltage follower may be interposed as a buffer between the peak detector and the load.

5-14 CURRENT-TO-VOLTAGE CONVERTER[8]

The current to be measured in the circuit of Fig. 5.11(a) is shown as being injected directly into the summing node (inverting terminal) of an OP AMP. This method of measuring a small current departs from the more conventional method of inserting a known resistor in the circuit and measuring the voltage developed across it. Another method for measuring such small currents might be that of using an OP AMP to amplify the small voltage across the inserted resistor; this, however, brings in the attendant error due to offset voltage.

The method of current-to-voltage conversion avoids these problems by making use of the basic OP-AMP property, where it forces the current through R_f continually to be equal to the input current I_{in}. The output voltage v_o thus simply becomes $-I_{in}(R_f)$.

An example of the use of the current-to-voltage conversion for a photoconductive cell is shown in part (b) of the figure. Let us assume that a change from dark current to illuminated current (ΔI_{in}) is 50 μA. With a value of 100 kΩ used for the feedback resistor (R_f), the voltage output change (Δv_o) will be

$$\Delta v_o = -50(10^{-6}) \text{ A} \times 100 \text{ k}\Omega = 5,000(10^{-3}) \text{ V}$$

$$= -5\text{-V change}$$

[8] *Linear Applications*, National Semiconductor Corp.; see address in Appendix IV.

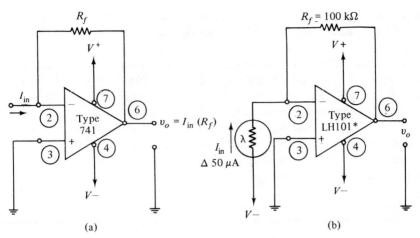

*LH101 is an internally compensated version of type LM101, and also equivalent to type 741.

Figure 5.11   Current-to-voltage converter: (a) general circuit; (b) converting a change in photocell current (ΔI_{in} = 50 μA) to a voltage change ($\Delta v_o = -\Delta I_{in}(R_f)$ = $-$5 volts). (*National Semiconductor*)

In the conversion process, the only error is the bias current (I_B), which is summed algebraically with the input current (I_{in}).

5-15 DIFFERENTIAL (INSTRUMENTATION-TYPE) AMPLIFIER

The requirements for an OP AMP used as an *instrumentation amplifier* are generally more rigid than those needed for simpler amplification purposes; in particular, the instrumentation amplifier must handle very small dc signal variations, and often these signals come through relatively long leads from remotely located sensors. This entails a combination of such properties as *high gain along with small offset errors and especially good common-mode rejection* properties that stretch the capabilities of general-purpose OP AMPs and often require the use of *premium monolithic types*, or possibly the use of the more expensive *hybrid modules* (discussed in Chapter 12). Where the requirements are not overly strict, however, the general-purpose OP AMP can be used in the *differential* (*or double-ended input*) *mode*, where the input is applied across the inverting and noninverting terminals, so that both input ends are "floating" above ground.

The connections for the differential-amplifier mode (not to be confused

with the diff-amp *stage* of Chapter 2) are shown in Fig. 5.12(a).[9] The signals
e_1 and e_2 usually come from a bridge arrangement of transducers [such as the
strain-gage amplifier in part (b) of the figure], and the signal to be amplified
represents the *difference* between the voltages at opposite corners of the bridge
$(e_1 - e_2)$. For this reason, the *dc differential amplifier* is also known by such
names as a *difference amplifier* or *error amplifier*. As previously noted, in the
differential connection, neither e_1 and e_2 is at ground potential.

In this differential-amplifier connection, with the OP AMP one retains the
valuable property of a high common-mode rejection ratio (CMRR) and also
keeps the offset errors very small by the selection of precision resistors to provide a proper balanced arrangement of the four resistors. The arrangement for
providing a good initial balance is obtained by using four well-matched resistors
to equalize the resistor ratios, as follows:

$$\frac{R_2}{R_1} = \frac{R_4}{R_3}.$$

In this way, the output of the differential amplifier responds only to the
difference of the input signals $(e_1 - e_2)$ and is thus able to cancel out common-
mode signals and also equal-and-opposite offset tendencies. As a result, the OP
AMP can amplify the signals from the unbalanced bridge with the least interference from undesired signals.

Bridge Amplifier

An example of the OP AMP in an instrumentation application is given for the
strain-gage amplifier of Fig. 5.12(b).[10] (This bridge type of instrumentation applies similarly to other sensors, such as thermistors, thermocouples, photocells,
and the like.) It will be noted in the circuit diagram that the bridge supply is
grounded, and that each signal voltage $(e_1$ and $e_2)$ is above ground by approximately the value of R. By using equal values for the feedback resistors (R_f) and
equal arms for the bridge (R), we satisfy the condition for good common-mode
rejection, since

$$\frac{R_2}{R_1} = \frac{R_4}{R_3}$$

becomes

$$\frac{R_f}{R} = \frac{R_f}{R}.$$

[9] Tobey and others, *Operational Amplifiers: Design and Applications.*
[10] Tobey and others, *Operational Amplifiers: Design and Applications.*

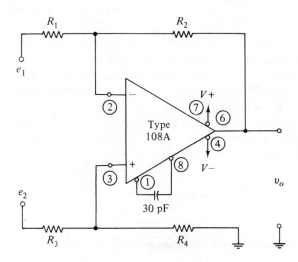

(a) Differential input to difference amplifier

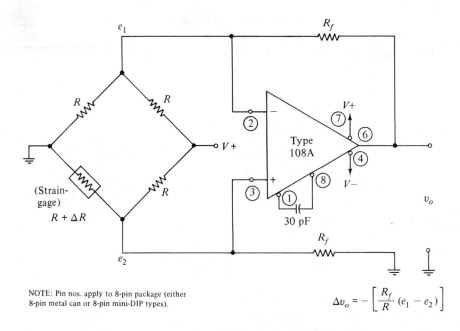

NOTE: Pin nos. apply to 8-pin package (either
8-pin metal can or 8-pin mini-DIP types).

$$\Delta v_o = - \left[\frac{R_f}{R} (e_1 - e_2) \right]$$

(b) Strain-gage bridge amplifier

Figure 5.12 Differential (double-ended input) amplifier: (a) double-ended input for amplification of difference $(e_1 - e_2)$; (b) example with strain-gage transducer bridge. (*National Semiconductor*)

Additionally, we retain initial equality of bias currents, since the resistance seen at each input terminal is the parallel combination $R_f \parallel R$.

The signal to be amplified comes from the change in resistance of the strain gage from its original value of R to a new value $(R + \Delta R)$. As an idea of the numerical values involved, we may assume an equal-arm-bridge value of $R = 100 \ \Omega$, and a change (ΔR) of $2 \ \Omega$, for a 2 percent change $(\delta = \Delta R/R = 2$ percent). With a supply voltage of +12 V, a Thèvenin-equivalent analysis[11] for the approximate bridge output gives

$$e_1 - e_2 = V \frac{\Delta R}{4R} \text{ V},$$

or
$$e_1 - e_2 = 12 \cdot \frac{2}{400} = 0.06 \text{ V} \quad \text{or} \quad 60 \text{ mV}.$$

To obtain an amplified output for this change in the range of volts, let us assume that a gain of 50 is needed. Since the approximate gain of this simplified circuit is $-R_f/R$, we could use the following values:

$$R_f = 5 \text{ k}\Omega, \quad R = 100 \ \Omega.$$

Then amplifier output v_o is

$$v_o = - \frac{R_f}{R} (e_1 - e_2),$$

and
$$\Delta v_o = - 50 \ (60 \text{ mV})$$
$$= 3\text{-V output change for a 2 percent change in } R.$$

In instances where unequal bias currents dictate the use of different values for the four resistors, the modified formula becomes more complicated, as follows:

$$v_o = - \left[\frac{R_2}{R_1} (e_1) \right] + \left[\left(\frac{R_4}{R_3 + R_4} \right) \left(\frac{R_1 + R_2}{R_1} \right) e_2 \right].$$

When it is desired to change the amount of gain, it is necessary to alter both ratios of the combination of resistors (rather than a single resistor) in order to preserve the balanced arrangement. (See more detailed discussion in Chapter 11.)

[11] Prensky, *Electronic Instrumentation.*

5-16 CURRENT-DIFFERENCE (NORTON) OP AMP: SINGLE-SUPPLY OPERATION

It is possible to operate the conventional dual-supply type of OP AMP with only a single supply; this can be done by the addition of two equal resistors in series as a bleeder across the single supply and arranged so that the resistors act as a voltage divider to supply half the voltage for each of the power input terminals of the OP AMP. However, for numerous *applications requiring only ac amplification* (as in audio and radio and television applications), a more practical alternative for single-power use is offered in the form of the *current-difference type of* OP AMP (not to be confused with the conventional type, usually called differential-input OP AMP). This current-differencing type operates on the *mirror-current* principle (previously described for the constant-current circuits of Figs. 2.16 and 3.4), where a fixed bias is applied to a given transistor to establish a constant current in it, and thus produce the same (or mirror) current for another transistor dependent on it.

The current-mirror action in the current-difference type of OP AMP is shown in Fig. 5.13(a), acting at the (+) input terminal for the noninverting function. When a positive voltage is applied to this (+) input terminal through an external resistor, the forward voltage drop across the diode produces a fixed bias for the current-mirror transistor in the lower corner, establishing a fixed value for its collector current, labeled I_{in+}.

This action is also shown in the diagram for the *inverting amplifier*, Fig. 5.13(b); here this action is seen at the noninverting (+) terminal, where $V+$ is applied to it, through the 2-MΩ resistor ($2R_2$). The resulting current (I) is shown to be effectively equal to the current in the feedback resistor (R_2), also labeled I. The values of external resistors (R_2 and $2R_2$) are chosen so that the quiescent dc output voltage, $V_{o(dc)}$ is $V+/2$, thus ensuring symmetrical positive and negative ac peaks.

The ac input signal is applied at the inverting terminal (−) through R_1 and produces another current (I_{in-}). It is the *difference between the two currents* (I_{in+} and I_{in-}) that is amplified and determines the ac output—from which it gets its name of *current-difference* amplifier.

In practice, the action of this inverting amplifier is very similar to the conventional OP AMP with dual-supply, as commonly used in a closed-loop. But here it is the *ac voltage gain* (A_v) that is determined by 'the ratio of the feedback resistor (R_2) to the input resistor (R_1): $-R_2/R_1 = -1\text{M}\Omega/100\text{k}\Omega = -10$ (in this case). Note also that the dc quiescent output of $V+/2$ volts caused by the positive $V+$ potential applied at the noninverting (+) terminal is blocked by the coupling capacitor used for ac amplification at both the input and output terminals.

For the *noninverting circuit* of the current-difference amplifier, shown in Fig. 5.13(c), the presence of a dc output of $V+/2$ volts also exists, with the

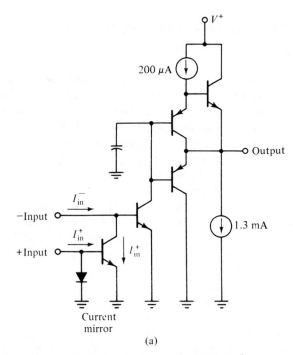

200 μA ↓

V^+

Output

I^-_{in}

−Input

I^+_{in}

+Input

I^+_{in}

1.3 mA

Current
mirror

(a)

Typical applications ($V^+ = 15\ V_{dc}$)

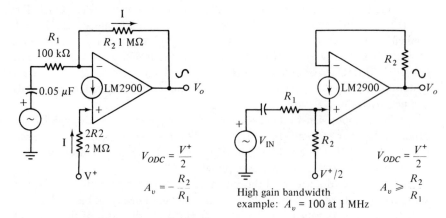

R_1
100 kΩ

I

R_2 1 MΩ

$-$

0.05 μF

LM2900

V_o

$+$

I

2R2
2 MΩ

V^+

$V_{ODC} = \dfrac{V^+}{2}$

$A_v = -\dfrac{R_2}{R_1}$

Inverting amplifier

(b)

R_2

$-$

LM2900

V_o

R_1

V_{IN}

$+$

R_2

$V^+/2$

High gain bandwidth
example: $A_v = 100$ at 1 MHz

$V_{ODC} = \dfrac{V^+}{2}$

$A_v \geqslant \dfrac{R_2}{R_1}$

Noninverting amplifier

(c)

Figure 5.13   The current-difference Norton type of OP AMP (*one quarter LM 3900*): (a) "current-mirror" circuit; (b) inverting and (c) non-inverting circuits, with current-source symbols to distinguish them from conventional type of OP AMP. (*National Semiconductor*)

in-phase ac voltage gain (A_v) approximately equal to the ratio

$$A_v = \frac{R_2}{R_1}$$

In this case, the positive voltage applied to the noninverting (+) terminal is only $V+/2$, but is applied through resistor R_2 which, in this case, is only half of what it was in the previous inverting configuration. Both of these circuits then are equivalent in establishing a dc quiescent value for the output that will allow symmetrical peak-to-peak output swings.

Each of the OP AMPs in Figs. 5.13(b) and (c) should be labeled to be *one quarter of LM 3900*, which is a quad OP AMP. Note also that each of the four units of the 3900 is not a pin-for-pin equivalent of a 741 type, as would be the case in a conventional quad OP AMP, such as a type 4741. In practice, however, the actual operation for each of these current-difference (also called *Norton*) OP AMPs is quite similar to the operation of a conventional one in obtaining a closed-loop gain by the ratio of the two external resistors (R_f/R_1), even though the internal method of operation differs. Essentially, the difference lies in the fact that the current-difference type amplifies a *difference of input currents*, while the conventional type amplifies a *difference of input voltages*. The main benefits of using the current-difference amplifier for ac amplification are associated with the use of an economical single supply (as in automotive circuits), and where the dc current drain from the supply remains practically independent of the value of the supply voltage. Additionally, the current-difference type lends itself more readily to the generation of waveforms up to around 1 MHz, with flexible output voltage swings having peak-to-peak values within 1 V of $V+$ (4 to 36 V). A large number of varied applications for the *quad LM 3900* are given in its data sheet.

True Differential-Input Voltage Quad

Another kind of single-supply quad is the *LM 324* quad OP AMP, which, however, operates on the more conventional principle of the difference between the input *voltages* (*rather than currents* as in the LM 3900). The 324 device is specially designed for single-supply operation; consequently, it is particularly suited for amplifying low-level *dc signals* that are referenced to ground (while the 3900 is more suitable for ac signals). In automotive applications, for instance, the 324 can handle dc signals from transducers such as thermocouples or strain gages, where these signals, for example, might feed into a digital microprocessor. Such signals require a zero output voltage for zero input, a feature conventionally obtained with dual supplies, but provided here for single-supply operation by a special design. The circuit of Fig. 5.14 illustrates the use of this

Typical single-supply applications (V^+ = 5 V dc)

Noninverting dc gain (0-V-input = 0-V output)

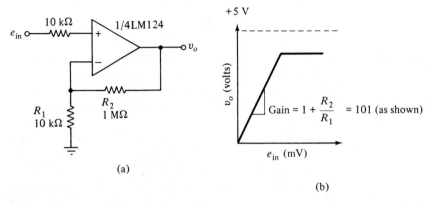

(a)

(b)

Figure 5.14 Noninverting amplifier using one quarter of quad LM 124/324 series for dc gain: (a) circuit; (b) output versus input graph. (*National Semiconductor*)

feature, using *one quarter of the LM 324*, and also shows a graph representing output versus input for dc or low-frequency transducer signals in this circuit.

Other examples of triple and quad OP AMPs are discussed in Chapter 12.

5-17 ADDITIONAL OP-AMP AND LIC CIRCUITS (BiFET AND BiMOS)

The dozen or so circuit diagrams given as examples in this chapter by no means exhaust the list of applications using OP AMPs. The examples given have been chosen to illustrate a fair sampling of typical OP AMP uses; additionally, however, we find OP AMPs used in many other circuits, where linear ICs are employed under specific names, and these other applications are covered later in separate chapters; they include such functions as *comparators, regulators, active filters, communication-type amplifiers, digital-interface devices,* and *hybrid IC modules* in specialized system applications.

While the field of OP-AMP applications continues to widen, it is well at this point to emphasize the unifying theme that the operational amplifier in the integrated-circuit form has evolved into a basic building block of great versatility. As a consequence, the OP AMP has presented the circuit designer with a revolutionary device for realizing applications that represent significant advances in both traditional and modern circuitry.

Monolithic Combination of Bipolar and Unipolar (FET) Transistors (BiFET and BiMOS)

The advanced fabrication technique, known as either BiFET or BiMOS, has made the combination of bipolar and FET devices on the same chip a practical development and has proved very popular. Overcoming the original drawback of an undesirably high input voltage offset (V_{io}), the newer method of incorporating the FET with the bipolar transistors provides a manageable V_{io} *offset figure*, while still retaining the obvious FET advantages of *very high input impedance* (R_{in}) along with very *low input-bias currents* (I_B). Together with various offset-trimming techniques (including *laser trimming*), the newer competitively priced combination BiFET and BiMOS devices provide improved performance characteristics in many cases over the older standardized ICs of the 741 and similar types.

An outstanding example of the improvement made by laser trimming is *National's LFT356 BiFET*, having an input offset (V_{io}) specified down to 0.5 mV, compared to the usual 5- to 6-mV value for the 741-type devices.

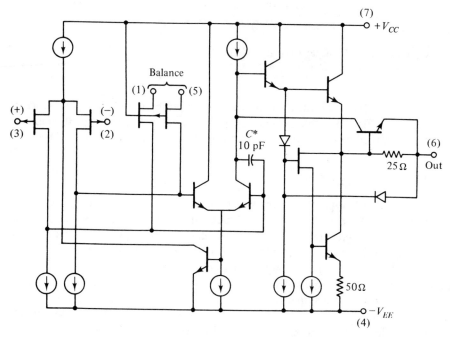

*C = 2 pF on LF157

Figure 5.15 Simplified schematic of a JFET (155/156/157 series) input OP AMP. (*Signetics*)

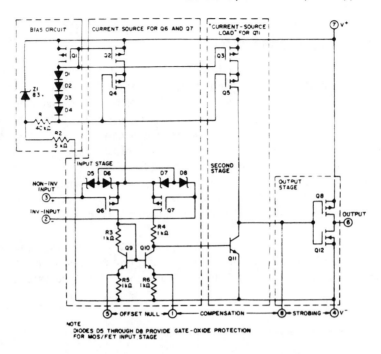

Figure 5.16 Schematic diagram of the CA3130 series BiMOS OP AMPs. (*RCA Solid State Division*)

Families in the BiFET field are exemplified by the 155/156/157 series by Signetics (Fig. 5.15) and the TL061/071/081 series by TI, while the BiMOS type is exemplified by CA3130/3140 series (Fig. 5.16) from RCA, among others.

BiFET versus BiMOS

In the BiFET type of device, which is produced by the majority of linear semiconductor manufacturers, the input pair of transistors are junction FETs (JFETs), as contrasted with the BiMOS device, which (with RCA as its main sponsor) uses the metallic oxide semiconductor (MOS) (or equivalently the insulated gate, IGFET) as the input transistor pair.

The BiMOS type (such as the *CA3130*) finds its major applications in single power-supply systems. With a single 5-V supply, for example, both types interface easily with 5-V logic levels; additionally, the BiMOS type exhibits slightly lower input current than the BiFET equivalent. On the other hand, the BiFET (such as the *LF356*) is superior with regard to noise, and when operated at ±15 V, it shows slightly better input-offset voltage characteristics than the CA3130.

Duals and Quads

The trend to provide dual and quad versions of various ICs is carried out in practically all the BiFET and BiMOS families. Examples of a progression of singles, duals, and quads are found in such families for BiFET in the LF351/353/347 progression from National and the TL081/082/084 progression from TI; also in BiMOS singles and duals form a progression of CA3140/3240 from RCA.

QUESTIONS

5-1. The main effect of using negative feedback in closed-loop operation causes
(a) Oscillation
(b) A higher gain than the open-loop gain
(c) A lower gain than the open-loop gain

5-2. For stable operation of an OP AMP in the noninverting mode, the output is fed back
(a) To the noninverting (+) terminal
(b) To the inverting (−) terminal
(c) To the ground terminal

5-3. To obtain operation of an OP AMP as a practical oscillator, use is made of
(a) Negative feedback
(b) Positive feedback
(c) Open-loop operation

5-4. The use of a *current-mirror* circuit provides
(a) Distortionless operation
(b) Higher-gain operation
(c) Constant-current operation

5-5. In operation as an *RC* oscillator, the OP AMP produces square waves rather than sine waves because of
(a) Saturation effect
(b) Long-time constant
(c) Absence of inductance

5-6. The action of an OP AMP as an integrator is functionally similar to
(a) A low-pass filter
(b) A high-pass filter
(c) A distortionless amplifier

5-7. An OP AMP operating as a summing amplifier can add
   (a) Voltages of only the same polarity
   (b) Voltages of only the same magnitude
   (c) Voltages of dissimilar polarity or magnitude

5-8. The half-wave rectifier of Fig. 5.9 operates in a more linear fashion because
   (a) The diode has a linear characteristic
   (b) The voltage required for diode conduction is negligible
   (c) Negative feedback provides a practically continuous amount of voltage required for diode conduction

5-9. Precision instrumentation amplifiers are generally made up of
   (a) Multiple OP AMPS
   (b) A single OP AMP
   (c) A single OP AMP with extra resistors

PROBLEMS

5-1. In the circuit of Fig. 5.1, assume $R_1 = 2.2$ kΩ and $R_f = 82$ kΩ. When +30 mV is applied to the *inverting* (−) *input*, find the output voltage (assume zero offset and nonsaturation of the amplifier).

5-2. Using the *noninverting circuit* of Fig. 5.2(a), with the gain setting $R_1 = 2.2$ kΩ and $R_f = 82$ kΩ, and making the same assumptions as in Problem 5-1, find the output voltage for this amplifier when a +30-mV signal is applied.

5-3. In the summing amplifier circuit of Fig. 5.4, assume $R_1 = 120$ kΩ and $R_2 = R_3 = 60$ kΩ.
   (a) Find the output voltage when $e_1 = +2$ V and $e_2 = e_3 = +1$ V.
   (b) What voltage must be applied to e_1 in part (a) to make the output approximately zero?

5-4. If the summing circuit of Fig. 5.4 remains as shown except for $R_f = 360$ kΩ
   (a) Find the new output for $e_1 = 2$ V and $e_2 = e_3 = 1$ V.
   (b) Would the assumption of nonsaturation be valid for a type 741 OP AMP in this case?

5-5. In the voltage-follower circuit of Fig. 5.3(a), sketch the output versus the input signal
   (a) When a 20-V p-p square wave is applied to the input
   (b) When a 20-V rms sine wave is applied to the input
   (c) Will the tendency to oscillate be greater with a type 741 or type 748 in this circuit? Explain.

5–6. Show what sort of circuit will produce quasi-voltage-follower action with a gain of 5 instead of unity.

5–7. In the integrating circuit of Fig. 5.5, $R_1 = 330$ kΩ and $C = 0.47\ \mu$F, with a dc input voltage of $+1.55$ V. Using the relation for dc, $V_o = -V_{in}\ [t/RC]$, how long does it take the ramp voltage to reach a level of -5 V (neglect the effect of R_2)?

5–8. Comparing the *noninverting mode of a conventional* OP AMP (Fig. 5.2) with the noninverting mode of a current-differencing type (Fig. 5.13), state the following:

    (a) Formula for the closed-loop gain (A_{VCL}) for each

    (b) Type of power supply used for each

    (c) Nature of the input voltage at the noninverting (+) terminal of each

    (d) Suitability of each for dc amplification

    (e) Nature of the input difference in each case that determines the output.

Chapter 6

Testing and Breadboarding Operational Amplifiers

6-1 LEVELS OF TESTING LINEAR INTEGRATED CIRCUITS

The measurements made in testing linear ICs will vary, according to the intended purpose, from relatively simple *manual tests on an individual IC*, as in the laboratory, to the much more complicated and extensive *automated test systems* used for production testing in industry. The test methods discussed here will concentrate on the OP AMP *as a typical linear IC*, with emphasis on the relative importance of the parameters to be measured. In this way those parameters of particular importance to other linear ICs can be identified and, with suitable modifications, can also serve to indicate tests for other groups of linear ICs. In addition, details of *practical means for setting up breadboard circuits* in the laboratory for performing these tests will be discussed.

6-2 BASIC LABORATORY TESTS

The chart of OP-AMP characteristics, given previously in Table 4-1, includes specifications that cover a wide range of properties that might enter into industrial uses. For testing in the laboratory, however, we list in Table 6-1 the relatively more important *basic properties for general-purpose use* (with symbols as shown in Fig. 6.1).

TABLE 6-1

Basic OP-AMP Properties for General Use

Open-Loop Properties	Closed-Loop Measurements
Voltage gain (A_{VOL} or A_o) Input resistance (R_{in}) Frequency-response curve: (f_c at -3 dB)	Large-signal gain (A_{VCL} or A_f) R_{in}, inverting mode R_{in}, noninverting mode Bandwidth (*BW*, at stated gain)

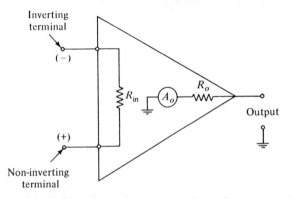

Figure 6.1 Simplified functional equivalent model of OP AMP: the open-loop gain (A_o or A_{VOL}), the input resistance (R_{in}) and output resistance (R_o) are intrinsic properties of the OP AMP, before any external components are connected.

Other properties, as applicable depending on circuit use, include input voltage offset (V_{io}), input bias current (I_B), voltage and current drift errors with temperature, output resistance (R_o), and effects of load resistance (R_L).

Test conditions, unless otherwise stated, are standardized as follows:

Supply voltage (dual) = ± 15 V,

Room temperature (T_A) = 25°C,

Load resistance (when used) ⩾ 2 kΩ.

6-3 VOLTAGE GAIN (TESTS FOR OPEN-LOOP, CLOSED-LOOP, AND LOOP GAIN)

Open-Loop Gain

A high value of the open-loop gain (A_{VOL} or A_o) is assumed in practically all the simplified formulas for the OP AMP. This assumption is validated in all the

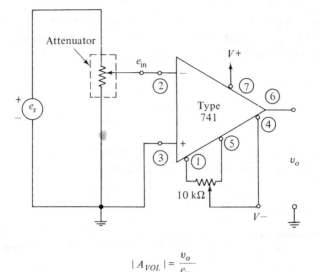

$$|A_{VOL}| = \frac{v_o}{e_{in}}$$

NOTE: Pin nos. apply to 8-pin package (either
8-pin metal can or 8-pin mini-DIP types).

Figure 6.2 Test for open-loop gain (A_{VOL} or A_o): control for *offset null* (10
kΩ) must be set for zero dc output at no-signal condition; e_s is an
ac signal, attenuated to avoid saturation of the OP AMP.

typical OP AMPs (as listed in Table 4-1), and is usually given in the data sheets,
where a plot of the *open-loop frequency response* shows values of A_{VOL} center-
ing around 100 dB (or 100,000 times) at the low and medium frequencies.

Because of the extremely high gain possible in open-loop operation, the
test for A_{VOL} is quite difficult to perform, and it is usually sufficient to take its
value from the frequency-response plot of the data sheet. The value may be
checked when desired, in the circuit of Fig. 6.2, by the magnitude relation

$$A_{VOL} = \frac{v_o}{e_{in}}$$

The use of the voltage divider for attenuation in Fig. 6.2 is made necessary by
the high gain and is used to ensure a sufficiently low input voltage (e_{in}) to avoid
saturation at the output. For similar reasons, a shielded input should be used, and
precautions must be taken to *null the initial dc input-offset voltage*, which
could cause amplifier saturation, even with zero ac voltage input.

Importance of Loop Gain (LG)

The specified value of open-loop gain for practically all OP AMPs is consistently
high enough to justify the simplified relation for the resulting gain in, for exam-

ple, an inverting amplifier, as the ratio of two external resistors (R_f/R_1). A parameter more important than A_{VOL} for determining the *accuracy of circuit gain*, however, is the *loop gain* (LG), which is the ratio of A_{VOL} to the actual closed-loop gain A_{VCL}, or

$$LG = \frac{A_{VOL}}{A_{VCL}} \quad \text{or} \quad \frac{A_o}{A_f}$$

This value depends on the amount of negative feedback used in the actual circuit, as discussed next for closed-loop operation.

Closed-Loop Gain (A_{VCL} or A_f)

The measurement of closed-loop gain in the circuit of Fig. 6.3 is the more practical measurement, since the OP AMP is very seldom used as open loop. For simplicity, the figure shows the circuit for the internally compensated type 741 connected for a nominal gain of 100. Under the conditions shown, the measure-

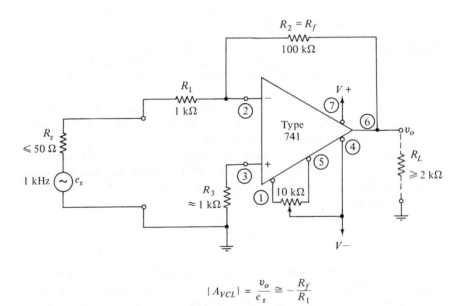

$$|A_{VCL}| = \frac{v_o}{e_s} \cong -\frac{R_f}{R_1}$$

NOTE: Pin nos. apply to 8-pin package (either
8-pin metal can or 8-pin mini-DIP types).

Figure 6.3 Test for closed-loop gain (A_{VCL} or A_f), for dc and frequencies within the audio band; for the actual gain to closely approximate $R_f/R_1 = 100$, R_1 must include R_s, and R_L must be sufficiently high to have negligible loading effect, as shown.

ment should closely approximate the simple relation

$$A_{VCL} \approx \frac{-R_f}{R_1}$$

The actual value obtained in this measurement is affected by two main approximations implied in this simplified circuit. First, the value used for R_1 should include the source resistance R_s. (In this case, the 50-Ω resistance of R_s may be neglected within 5 percent; otherwise, we either include the known resistance of the generator or, if unknown, we can use a voltage divider across the source to reduce the equivalent internal resistance to a 50-Ω value.) Second, the value of R_3 at the noninverting terminal should be the parallel combination $R_f \| R_1$ to equalize the bias currents at each terminal; here the round value of 1 kΩ for R_3 is within 1 percent and voltage offsets are assumed to be negligible for large-signal operation. In a similar way, the effect of using a load resistor (R_L) equal to or greater than 2 kΩ will not be appreciable, compared to the no-load condition.

Frequency-response curves at a desired gain within the audio range can also be obtained from the circuit of Fig. 6.3.

Example 1

Loop-Gain Calculation

Since the value of a loop gain (LG = A_o/A_f) has important significance in determining accuracy and stability considerations, it is well to evaluate it for a given circuit, once the closed-loop gain (A_{VCL} or A_f) is found. Using a typical value for the open-loop gain (A_{VOL} or A_o) of 100 dB, or 100,000 times, we can easily find the loop-gain value for the circuit of Fig. 6.3, where A_{VCL} or A_f = 100 or 40 dB, as follows:

$$LG = \frac{A_o}{A_f}$$

$$= 100 - 40 = 60 \text{ dB, or}$$

$$= \frac{100,000}{100} = 1,000 \text{ times.}$$

This value of 1,000 (or 60 dB) for the loop-gain indicates the gain-reduction factor (sometimes called the "throwaway" factor) introduced by the negative feedback.

It should be noted that the loop gain obtained in this fashion is the same gain-reduction factor that comes from the familiar generalized feedback relation

$$A_f = \frac{A_o}{1 + A_o \beta},$$

where β is the fraction of the output that is returned as negative feedback. Thus the loop gain is the expression in the denominator $1 + A_o \beta$, simplified by neglecting to add the 1, or

$$LG \cong A_o \beta$$

$$= 100,000 \left(\frac{R_1}{R_1 + R_f} \right)$$

$$= 100,000 \, \frac{1 \text{ k}\Omega}{101 \text{ k}\Omega}$$

$$\cong 1,000, \quad \text{as before.}$$

A large value of loop-gain is generally advisable, since it can be shown that it reduces, by the same factor, the effect of the possible spread of parameter variations from one device to another. Thus the *value of loop-gain is significant in determining the expected accuracy of the closed-loop gain.* It is also a significant multiplying factor in obtaining a high input impedance in the noninverting mode, as will be seen in the next section, where both the gain and input impedance for the noninverting mode are discussed.

6-4 INPUT-RESISTANCE TESTS

Open-Loop (R_{in}, Intrinsic)

The property of input resistance (R_{in}) of an OP AMP under open-loop conditions is intrinsic to the device and accordingly is shown on the data sheet. It can be checked by the circuit of Fig. 6.4.

Values of R_{in} encountered in typical OP AMPs are much greater than the few thousand ohms that might be expected from the conventional input resistance ($2h_{ie}$) of an ordinary differential-amplifier input stage.

Again taking the internally compensated type 741 as our example, for the sake of simplicity, we find that the value of the intrinsic R_{in} ranges from 300 kΩ to the megohm range, this relatively high value being obtained by the use of a modified Darlington circuit at the input. Another scheme for increasing R_{in} is the use of superbeta transistors, as in the type LM108A; such specially fabricated transistors can provide a beta of around 5,000 at only 1 μA of collector

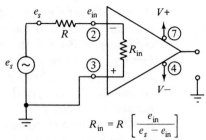

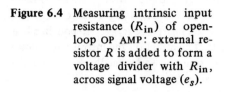

Figure 6.4 Measuring intrinsic input resistance (R_{in}) of open-loop OP AMP: external resistor R is added to form a voltage divider with R_{in}, across signal voltage (e_s).

$$R_{in} = R \left[\frac{e_{in}}{e_s - e_{in}} \right]$$

NOTE: Pin nos. apply to 8-pin package (either 8-pin metal can or 8-pin mini-DIP types).

current, producing a resulting R_{in} of around 30 MΩ. And where still higher values for R_{in} are important, the example of the type 770 is given in the OP-AMP chart (Table 4-1) as providing up to 100 MΩ.

Still larger values are available with the use of special-purpose OP AMPS with FET inputs (such as type 740), providing a listed value of 1 million MΩ (10^{12} Ω).

Reverting to the type 741 of Fig. 6.4 for simple laboratory testing, R_{in} is measured as the differential resistance of the OP AMP by inserting a high resistance (R) in series with the input terminal to act as a voltage divider. (Note that for open-loop conditions only, either input terminal may be used.) As a result, the signal voltage e_s divides, with the fraction e_{in} appearing across R_{in} as

$$e_{in} = e_s \frac{R_{in}}{R + R_{in}} .$$

Solving for R_{in}, gives

$$R_{in} = R \frac{e_{in}}{e_s - e_{in}} .$$

(Note that if R is a variable resistor and is adjusted to make $e_{in} = \frac{1}{2} e_s$, then $R_{in} = R$.) For OP AMPs having an FET input stage, the preceding method is not practical because $e_s \approx e_{in}$.

Closed-Loop R_{in}

1. *Inverting mode.* When the OP AMP is used closed-loop in the inverting mode (as was shown in Fig. 6.3), the *input resistance of the circuit is radically reduced by the loop gain* (as in shunt voltage feedback), so that for closed-loop the approximate expression becomes

$$R_{in} \text{ (for inverting model)} \approx R_1$$

2. *High-impedance noninverting mode.* When the input to the OP AMP is connected to the noninverting (+) terminal under the closed-loop conditions of Fig. 6.5, the *input impedance of the circuit is radically increased by the loop-gain factor.* [The gain, however, is substantially the same, changing only slightly to $(1 + R_f/R_1)$.] Resistor R_{eq} (including R_s) should equal the parallel resistance $R_f \| R_1$ to equalize bias currents for minimum dc offsets. The effect on the intrinsic resistance is now similar to changing from shunt to series feedback, so that the expression for the new input resistance is

$$R_{in} \text{ (for noninverting mode)} = R_{in\,(\text{intrinsic})} \text{ (LG)}.$$

Since loop gain (LG) is defined as

$$\text{LG} = \frac{A_o}{A_f} \quad (\text{or } A_o\beta),$$

then R_{in} (for noninverting mode) $= R_{in\,(\text{intrinsic})} \dfrac{A_o}{1 + \dfrac{R_f}{R_1}}$.

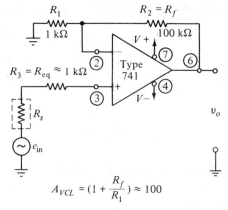

$$A_{VCL} = (1 + \frac{R_f}{R_1}) \approx 100$$

R_{in} (non-inverting mode) $= R_{in}$ (intrinsic) (Loop gain)
$$\approx 300 \text{ k}\Omega \text{ (1,000)} = 300 \text{ M}\Omega$$

NOTE: Pin nos. apply to 8-pin package (either 8-pin metal can or 8-pin mini-DIP types).

Figure 6.5   High input resistance (R_{in}) for the noninverting mode of OP AMP operation: while the closed-loop gain (A_{VCL}) of this circuit is roughly the same as in the inverting mode of Fig. 6.3, the input resistance (R_{in}) has increased from 1 kΩ to about 300 MΩ. (*Fairchild Camera and Instrument Corporation*)

Using the example of *loop gain of 1,000* (given in Section 6-3 for a closed-loop gain of 100), we can expect a multiplying factor of 1,000 times the intrinsic R_{in} (given as a minimum of 300 kΩ on the data sheet for the type 741).[1] This calculation gives a theoretical value of 300 MΩ in the noninverting mode, which agrees well with the stated value of 280 MΩ for R_{in} (as given on this data sheet for a closed-loop gain of 100).

6-5 BANDWIDTH (ESTIMATING AND TESTING)

The open-loop frequency response of an OP AMP is given in its data sheet and is shown for a typical case in Fig. 6.6(b). The type 101A illustrated (similar to the 748 type) is known as an *extended bandwidth* OP AMP. Here the bandwidth (*BW*) is made more flexible by allowing the use of different values for the external compensating capacitor (C_1), rather than the situation where the bandwidth is restricted in the internally compensated type (such as type 741).

Although the OP AMP is rarely used open-loop, we can *easily estimate the closed-loop performance from the open-loop response curve* by employing the concept of constant *gain* × *BW* product. Thus, in the plot of Fig. 6.6(b), we estimate the bandwidth for a closed-loop gain of 100 (or 40 dB) by simply moving horizontally to the first curve (C_1 = 30 pF), and finding an approximate *breakpoint* (*or* f_c) *at 10 kHz*, after which the gain rolls off at 6 dB octave (or 20 dB decade) for increasing frequency. This form of roll-off is known as *single-pole compensation.* (The circuit for two-pole compensation is also generally given in the data sheets.)

The value of 30 pF provides adequate compensation (or protection against oscillation) for all gains down to unity gain, where the quoted amount of compensating capacitance is needed. For larger gains, smaller amounts of compensating capacitance are satisfactory and the resultant bandwidth is "extended." Thus, at the same gain of 40 dB (100 times), the horizontal extension to the curve for C_1 = 3 pF intersects at 100 kHz, indicating a *tenfold increase in bandwidth*. A point of interest in this connection is the fact that in choosing an OP AMP for fairly high gain in the complete audio-frequency range, 748, the 101A, or similar types are preferred over the internally compensated 741 types, whose bandwidth is similar to the 30-pF curve (*BW* = 10 kHz at a gain of 100).

Test Circuit

The diagram [Fig. 6.6(a)] shown for testing the bandwidth (or frequency response) in closed loop is tailored for the desired gain by the ratio R_f/R_1 (thus R_f = 1 MΩ for a gain of 100). The input resistor R_1 is given arbitrarily as 10 kΩ

---

[1] Fairchild μA741 data sheet.

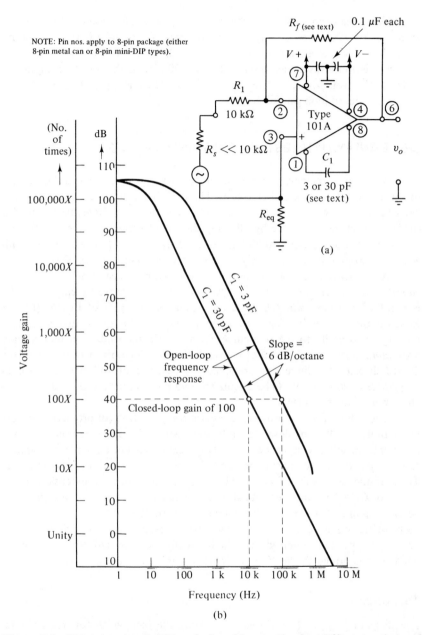

Figure 6.6 Obtaining bandwidth of closed-loop gain for different values of frequency-compensating capacitor C_1: (a) for circuit gain of 100, $R_f = 1$ MΩ, $R_{eq} = R_f \| R_1$; (b) two open-loop frequency response curves. (*National Semiconductor*)

to allow for the signal-source resistance R_s, since the published curves apply only if R_s is less than 10 kΩ. The value for R_{eq} is the equivalent resistance of the parallel combination $R_f \| R_1$ (a 10-kΩ resistor would be satisfactory here with most signal generators). The 0.1-μF capacitors are not critical and are used to bypass the positive and negative power supplies.

The bandwidth of the circuit is tested by increasing the input frequency of a signal source whose amplitude is constant for all frequencies to *the point where the output drops to 70.7 percent (–3 dB) of the low-frequency value.* The bandwidth would then extend from direct current to this point. However, if the dc gain is to be measured, the offset-null circuit must be included to prevent the amplified input-voltage offset from appearing in the output.

It should be kept in mind that wide-band amplifiers are available as special linear ICs to provide much wider bandwidths than the general-purpose OP AMPs discussed here; they are covered in a separate chapter.

6-6 OUTPUT-RESISTANCE (R_o) TEST

The output-resistance specification for an OP AMP is generally of little importance in voltage amplification, especially when reasonably high load resistances (>2 to 10 kΩ) are used. It assumes greater importance when relatively high current outputs are required, as in booster amplifiers, which are discussed separately later.

The following discussion indicates the nominal values of R_o to be expected under the three operating conditions of open-loop and of inverting and noninverting modes of closed-loop operation.

Open-Loop R_o

The measurement of intrinsic output resistance (open-loop) follows the same plan of using a voltage-divider circuit as was used previously for the intrinsic input-resistance test. In this case, as seen in the equivalent model of Fig. 6.7, load resistance (R_L) acts to reduce the original open-loop output voltage (V_{OL}) before loading to a smaller value (v_o) after R_L is connected. Again here, as in previous open-loop measurements, the difficulty in working with the very high open-loop gain suggests the advisability of using the value of R_o given in the data sheet as an initial figure for specific calculations. This value centers around 100 Ω.

When it is desired to check the value of R_o in open loop, the voltage divider relation shown in Fig. 6.7 can be used as follows:

$$v_o = V_{OL} \frac{R_L}{R_o + R_L}.$$

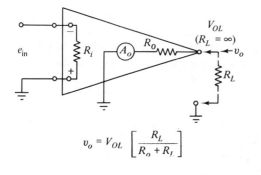

$$v_o = V_{OL}\left[\frac{R_L}{R_o + R_I}\right]$$

Figure 6.7 Testing intrinsic output resistance (R_o): the unloaded open-loop output voltage (V_{OL}) is reduced to v_o, when the load resistor (R_L) is connected.

from which
$$R_o = R_L\frac{V_{OL} - v_o}{v_o}.$$

(Note that when v_o is made equal to $\frac{1}{2}V_{OL}$, then $R_o = R_L$.)

Closed-Loop R_o

The output resistance is reduced by negative feedback in both the inverting and noninverting modes. Consequently, the expression for R_o is identical in both configurations:

$$R_{o(\text{closed loop})} = \frac{R_{o(\text{intrinsic})}}{LG}$$

$$= R_{o(\text{intrinsic})}\frac{A_f}{A_o}.$$

For the example used before of loop gain = 1,000, we find the satisfyingly small value of R_o to be around $100/1,000$, or $0.1\ \Omega$.

6-7 COMMON-MODE REJECTION TESTS

An outstanding advantage of the theoretical operational amplifier is that there is *zero output for common-mode signals* (i.e., for equal signals on both inputs to the differential-amplifier stage). This theoretical infinite rejection is modified by the practical consideration that both sides of the input stage cannot be expected to be exactly identical, and so the practical common-mode rejection ratio (CMRR), as a measure of the actual unbalances, is defined as a ratio, comparing differential gain (A_d), that is, the gain to the difference signal $(e_1 - e_2)$, to the gain for common-mode signals (A_{cm}), or

$$\text{CMRR} = \frac{A_d}{A_{cm}}.$$

Thus, particularly in noisy locations, a large value of this ratio is desired within the limits of the magnitude of allowable common-mode voltages that are specified.

In the CMRR test circuit of Fig. 6.8, the signal voltage (e_s) is essentially the common-mode voltage (v_c) applied simultaneously to both of the OP AMP input terminals. Under the conditions of matched resistors, $R_{f1} = R_{f2} = R_f$ and $R_1 = R_2 = R$, the relation for the common-mode rejection becomes

$$\text{CMRR} = \frac{A_d}{A_{cm}} = \frac{R + R_f}{R} \frac{e_s}{v_o}.$$

For a numerical example, let $R = 100 \ \Omega$ and $R_f = 100 \ \text{k}\Omega$, for a nominal gain of 1,000. If a signal voltage e_s of 1 V is applied and produces an output v_o of 10 mV, then

$$\text{CMRR} = \frac{100 + 100,000}{100} \frac{1,000 \ \text{mV}}{10 \ \text{mV}}$$

$$\approx 1,000 \ (100)$$

$$= 100,000 \ \text{times} \quad \text{or} \quad 100 \ \text{dB}.$$

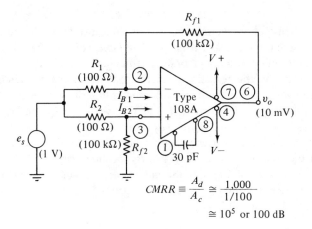

$$CMRR \equiv \frac{A_d}{A_c} \cong \frac{1,000}{1/100}$$

$$\cong 10^5 \text{ or } 100 \text{ dB}$$

NOTE: Pin nos. apply to 8-pin package (either 8-pin metal can or 8-pin mini-DIP types).

Figure 6.8 Measuring common-mode rejection ratio (CMRR): a 1-V ac signal applied equally to both input terminals yields an output v_o of 10 mV ($A_{cm} = 1/100$), while the circuit has a differential gain (A_d) of 1,000, resulting in a ratio of 10^5 or 100 dB for CMRR.

This value of 100 dB is fairly typical of precision OP AMPs (such as type μA725 or type LM108A), providing the measurement is made within the specified common-voltage swing.

6-8 INPUT-OFFSET ERROR (I_B, I_{io}, AND V_{io}) TESTS

The factors contributing to the unbalances that give rise to offset errors may be summarized as five in number, each perhaps negligibly small, but potentially significant, depending on the application. These factors are listed in Table 6-2 with typical magnitudes for two types: the internally compensated 741 type and the superbeta 108A type, as given in Table 4-1.

TABLE 6-2
Offset Factors

	Type 741	Type 108A
Input bias current (I_B)	<200 nA	<2 nA
Input offsct current (I_{io})	<30 nA	<0.2 nA
Drift of I_{io}	<200 pA/°C	<2.5 pA/°C
Input offset voltage (V_{io})	<5 mV	<1 mV
Drift of V_{io}	20 μV/°C	5 μV/°C

In this list of offset factors, we can concentrate on the ones most commonly measured, *input bias current I_B*, and *input voltage offset V_{io}*, since these are dominant under ordinary laboratory temperature conditions.

Accordingly, the test circuit of Fig. 6.9 shows the measurement for the input offset voltage (V_{io}); also, placing a meter in series with each input terminal will determine input bias current (I_B) as the average of I_{B1} and I_{B2}, and determine I_{io} as their difference ($I_{B1} - I_{B2}$). But the effect of I_B or I_{io} is generally negligible, assuming significance only when large feedback resistors are used such that the voltage drop $\Delta I_B R_f$ (or $I_{io}R_f$) becomes large.

In the circuit of Fig. 6.9 the unbalance producing the offset voltage at the input is amplified by the closed-circuit gain of the OP AMP, so that the value of V_{io} is obtained as follows:

$$V_{io} = \frac{v_o}{A_{VCL}} .$$

For a numerical example using the closed-loop gain of 100 in the figure and the typical offset values mentioned previously, we have the data shown in Example 2.

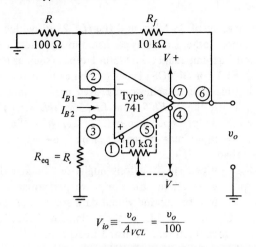

$$V_{io} \equiv \frac{v_o}{A_{VCL}} = \frac{v_o}{100}$$

NOTE: Pin nos. apply to 8-pin package (either
8-pin metal can or 8-pin mini-DIP types).

Figure 6.9 Measuring input offset voltage (V_{io}, also called V_{os}): $I_B = (I_{B1} + I_{B2})/2$, and $I_{io} = I_{B1} - I_{B2}$ may be measured, as described in the text. The dashed connections are for offset null control, when used to cancel V_{io} in dc applications.

Example 2

	Type 741	Type 108A
$V_{io} = \dfrac{v_o}{A_{VCL}}$	$= \dfrac{500\,\text{mV}}{100} = 5\,\text{mV}$	$= \dfrac{100\,\text{mV}}{100} = 1\,\text{mV}$
Contribution of $I_{io} R_f$	$= 30\,\text{nA}\,(10{,}000) = 0.3\,\text{mV}$	$= 0.2\,\text{nA}\,(10{,}000) = 2\,\mu\text{V}$

For general-purpose dc applications (such as with the type 741 OP AMP), the *offset-null provision* allows cancellation of the initial offset error by manual adjustment.

6-9 IMPROVED SPECIFICATION TYPES OF OP AMPS

In addition to the "standardized" types of general-purpose OP AMPs, there are the newer versions with *improved specifications*, forming a group that fits in between the older general-purpose types discussed in previous paragraphs and the specialized precision instrumentation group that is covered in Chapter 11. Typical examples in this possibly overlapping group are discussed next.

For the popular general-purpose 709/741 types, we find improved specifications for wide-band operation in types such as the *CA 3100*, having a unity-gain crossover frequency of 38 MHz, and the *HA 2530*, which has a high slew rate of 320 V/μs, among other similar types listed in Appendix III.

For improved high-impedance FET-input types (such as the 740), a newer development of Bi-FET (or Bi-MOS) types combines both the bipolar and FET technology and is thus able to achieve improved performance along with high input impedance. Thus, the *Bi-FET type LF156A* (*National*) combines wide bandwidth with a high input impedance of 10^{12} Ω, while the *Bi-MOS type CA 3130* combines a low input bias current (I_B) of 5 pA with a low-drift input offset (V_{io}) of less than 5 μV/°C.

While all these refined specifications might not be strictly necessary for many general-purpose applications, they improve performance in many cases. The specifications, as given on the individual data sheets, are carefully tested in industrial production. Other examples of these "improved-spec" general-purpose devices from various manufacturers are listed in Appendix III and the addresses for obtaining data sheets are in Appendix IV.

6-10 CURVE-TRACER METHOD OF TESTING AMPLIFIERS

A fast and convenient method for determining satisfactory operating conditions of an OP AMP makes use of a curve-tracer, where the transfer function of the device under test (DUT) is displayed on an oscilloscope. This method involves using a sawtooth signal to sweep the input voltage through a stated range of values, and applying the output to the vertical terminal of an oscilloscope, while the same sawtooth is connected to the horizontal terminal. The resulting trace is an input–output graph of the transfer function on the scope screen. The trace can be used as a visual GO/NO-GO indication, but, more importantly, it also provides speedy information on three main OP-AMP characteristis; *gain* (either A_{VOL} or A_{VCL}), *input voltage offset* V_{io}, and *maximum output voltage limits*.

The details of the test method are very similar to those for a simple diode curve tracer, with the active OP AMP taking the place of the diode being tested. The circuit requirements for the tester (shown in Fig. 6.10) are even simpler than the circuit needed for a transistor curve tracer since, as a two-port device, the OP AMP does not require a stepping voltage. The OP-AMP tester circuit of Fig. 6.10 was developed[2] to give the user "a degree of confidence that the units are functioning satisfactorily."

The drive circuit, in parts (a) and (b) of the figure, uses a portion of the 60-Hz voltage (from the 12-V ac secondary winding of the power transformer) and applies it (at point *A*) to the normally off transistor. This allows the *RC*

---

[2] "An Operational Amplifier Tester," Application Note AN 400, Motorola Semiconductor Products, Inc.

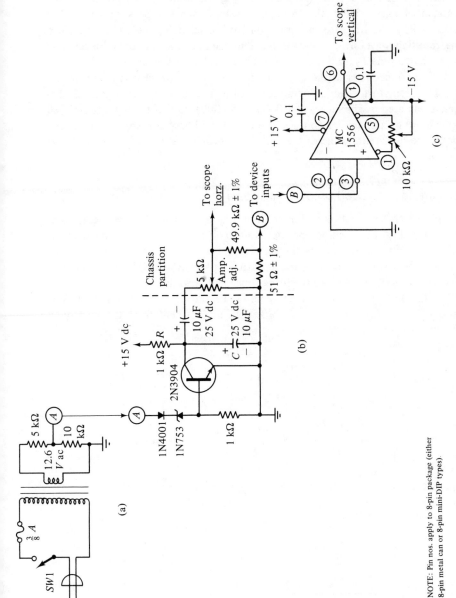

Figure 6.10 OP AMP tester using curve-tracer method: (a) sine-wave input to drive circuit ; (b) drive circuit producing ramp with fast retrace; (c) connection to OP AMP (DUT). (*Motorola Semiconductor Products, Inc.*)

combination to freely charge toward the +15-V dc power supply. With a sufficiently long time constant (relative to the 60-Hz supply), the charging curve reaches a sufficient amplitude, while deleting the more curved portion of the exponential curve. When the input sine wave reaches a peak of about 7 V $(2V_{BE} + V_Z)$, the transistor is turned on hard, discharging the timing capacitor. The resulting waveform, as shown, has the charging time T_1 much longer than the discharge time T_2, and thus the fast retrace is barely visible.

The connections to the device under test are shown in part (c) of the figure. The input to the noninverting terminal of the OP AMP (from point B) has been attenuated by the factor of 1,000 (51/50,000), so that the highly amplified (open-loop) output applied to the scope vertical will have a reasonable value relative to the horizontal value of the same waveform, as set by the 5-kΩ AMP ADJ control.

Interpreting the Oscilloscope Waveform

A typical transfer function is illustrated in Fig. 6.11, with the main features as noted. Initially, the oscilloscope inputs are grounded. The oscilloscope dot is then centered; this establishes the vertical and horizontal references. After the device to be tested has been connected, turn up the AMP ADJ until the unit is in deep saturation. Read directly positive and negative $v_{o(max)}$. The open-loop gain (A_{VOL}) is the calculated slope of the transfer curve times 1,000. The input offset voltage (V_{io}) is the horizontal displacement from the horizontal reference to the point where the function crosses the vertical reference divided

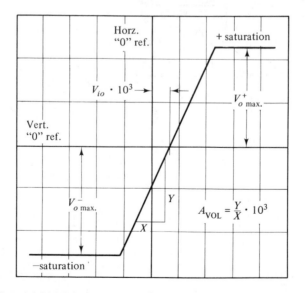

Figure 6.11 Interpreting oscilloscope trace of IC curve tracer.

by 1,000. The straight line sloping between the saturation points is also indica-tive of the device's linearity. In reading V_{io} and A_{VOL} from the scope presenta-tion, it is advisable to increase the horizontal sensitivity, so that the resolution is increased for better accuracy.

This kind of tester, when constructed with appropriate input jigs, forms a very handy do-it-yourself instrument in the laboratory for quickly resolving any doubts concerning a questionable OP AMP device.

Commercial Form of Curve-Tracer Testing

A commercial type of an LIC tester, producing a number of *oscilloscope-trace displays*, is offered as *test fixture 178*, which attaches to the Tektronix 577 Curve Tracer. With this connection, the curve-tracer measurement system adds to its conventional diode-transistor test capabilities, and provides for linear IC testing with a storage-scope (D_1) display, which accommodates the very slow sweep rate that must be used to display many LIC characteristics.

The typical characteristics that can be checked on an LIC amplifier with this curve-tracer system include open-loop gain (A_{VOL}), common-mode rejec-tion ratio (CMRR), power-supply rejection ratio (PSRR), input and supply cur-rents, along with checks on thermal drifts and noise.

Plug-in test cards are offered for defining the type of LIC that can be tested, one for testing IC amplifiers and the other for testing IC regulators. Each card can be quickly set up by jumper leads for the particular linear IC under test.[3]

6-11 BREADBOARDING THE INTEGRATED-CIRCUIT PACKAGE

The IC package comes in three main forms, as shown in Fig. 6.12: one form is the *can type* in part (a) (usually designated as TO-5 or TO-99 types), having 8, 10, or 12 leads; another main form is the *dual-in-line plastic* (DIP) *type*, having 14 or 16 leads in its full form in part (b); a variation of the DIP is the increas-ingly popular *Mini-DIP package* with 8 leads, shown in Fig. 6.12(c).

In commercial assembly practice, the accepted method for mounting the IC package is on a prepared printed-circuit (PC) board, and making the connec-tions presents no particular problem using commercial printed-circuit tech-niques. But for laboratory purposes there may be considerable difficulty in the use of commercial sockets, calling for connecting the many closely spaced leads (from 8 up to 16 leads) with the attendant likelihood of possible short circuits between adjacent leads.

[3] *Tekscope*, Nov. 6 issue from Textronix, P.O. Box 500, Beaverton, Ore. 97005.

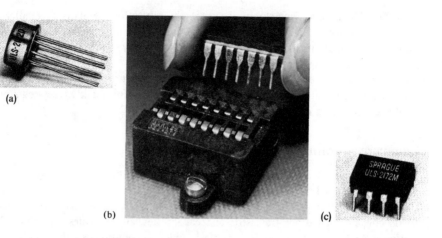

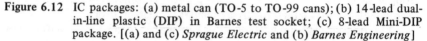

Figure 6.12 IC packages: (a) metal can (TO-5 to TO-99 cans); (b) 14-lead dual-in-line plastic (DIP) in Barnes test socket; (c) 8-lead Mini-DIP package. [(a) and (c) *Sprague Electric* and (b) *Barnes Engineering*]

Of the many possible ways for connecting the IC on an experimental breadboard, *laboratory practice strongly prefers a plug-in socket arrangement with the multiple leads brought out to numerous posts.* It is preferable also to allow for easy changeover of circuit configurations, wherever feasible. For this purpose, a perforated board is generally used, and Fig. 6.13 (a) shows one such arrangement, using a *Vector board*[4] with associated spring-clip terminal posts. When making connections in the laboratory to the multiple pins of the IC package, a number of special methods are available for avoiding the difficulty of soldering the closely spaced leads. *Sockets with long wire-wrap leads*, which may then be bent to spread out the spaces between the leads, are one suggestion; another is the use of *adhesive-backed conductors*, which have a printed conductive pattern extending the socket-pin contacts to more widely spaced terminals.

An effective alternative to the regular socket is the use of *multiple-terminal connection strips*, which, if desired, require no soldering at all. Such a multiple-terminal connection board, as shown in Fig. 6.13(b), acts as a socket to accept both DIP and TO-5 type packages, and simultaneously provides multiple terminals for each pin of the package. This is done by simply inserting the bared end of an insulated wire into any one of four holes, whose contacts are wired in parallel with each pin connection, thus providing for up to four additional connections (without soldering) to each of the package leads (EL Socket, types SK-10, 20, or 50).[5]

[4] Vector Electronics, 12460 Gladstone Ave., Sylmar, Calif. 91342.

[5] EL Instruments, Inc., 61 First St., Derby, Conn. 06418, who also supply a kit for a "Universal OP-AMP Breadbox."

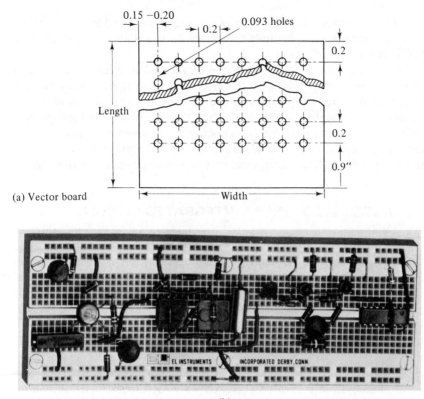

(a) Vector board

(b)

Figure 6.13 Convenient (no-soldering) socket and breadboard arrangement: (a) Vector perf-board (0.093 holes) takes Vector spring clips; (b) EL Instruments, socket SK-10 accepts DIP packages (and metal cans with formed leads) and provides four parallel-connected receptacles for each lead.

The breadboard multiterminal, illustrated in Fig. 6.13(b) is of the SK-10 type, which is long enough to accept eight of the 14-pin DIPs (or the equivalent in the 16-pin to 40-pin models). Shorter types, such as the SK-50, accept four of the 14-pin models, while the SK-20 is a basic unit, accepting just one 14-pin or 16-pin DIP. Since there are four receptacles for connecting wires to each pin, the multiterminal socket is convenient in providing all necessary input, output, and power-supply connections; moreover, it is generally possible to accommodate practically all external components (such as $\frac{1}{4}$-W resistors and capacitors with wire leads, number 22 or thinner), by plugging them directly into the receptacles.

Such a. multiple-terminal block allows easy substitution of components and, mounted on a perf-board with Vector spring clips that fit into the perf-board holes, combines to make a highly convenient and flexible type of bread-board. (A similar type of multiple-terminal block is manufactured by AP, Inc., and CS Corp.).[6]

In practical use, as in the illustrated breadboard arrangement, it is a very simple matter (without soldering) to make such circuit alterations as *changing the gain* of the circuit (by substituting a new resistor for R_f), *changing from inverting to noninverting mode* (by shifting the signal-source and ground con-nection), and even *changing from an amplifier to an oscillator circuit* (by simply substituting a capacitor for a single resistor). It thus offers a particularly useful and convenient breadboard arrangement for experimental use in the laboratory.

6-12 AUTOMATED LINEAR-INTEGRATED-CIRCUIT TESTING

The testing of linear ICs in quantity, as in production testing, calls for a tester capable of speedy and automatic operation. The industrial testing instruments range from the *semiautomatic types*, for which both manual and automatic modes are provided, to the highly developed *computer-controlled automatic types*.

An example of a highly versatile and speedy tester, which though auto-matic does not require connection to a computer, is illustrated in Fig. 6.14, the *GenRad GR 1731 linear tester*. It can be set up to provide as much informa-tion as desired, from a simple GO/NO-GO indication to a detailed account of each test, displaying any or all test results on a video or LED display.

It tests a wide variety of linear ICs, including not only OP AMPs, but also comparators, regulators, and other such low-voltage linear IC devices. Some of the parameters tested are listed in Table 6-3.

TABLE 6-3

Test Parameters for Automated Testing

Gain–bandwidth product ($G \times BW$)	Input offset voltage (V_{io})
Gain (with and without load) (A_v)	Input offset current (I_{io})
Common-mode limits ($\pm E_{cm}$)	Power-supply rejection ratio (PSRR)
Output impedance (Z_o)	Maximum output [V_o(max)]
Bias current (I_B)	Quiescent operating current ($\pm I_{CC}$)

Circuits can be tested as fast as they are connected to the device-adaptor boards, and one button initiates all tests once the desired conditions have been established. The proper conditions are set on interchangeable data cartridges for the selection of all tests; also skip tests and skip limits are included to make

[6] AP, Inc., 72 Corwin Drive, Painesville, Ohio 44077, and CS Corp., 70 Fulton Terrace, New Haven, Conn. 06509.

Figure 6.14 The 1731 Linear IC Test System in use with a handler which allows automatic binning of linear devices. (*GenRad*)

the tests as comprehensive or simple as desired. Together with various options as, for example, printing the results, the tester provides a wide range of tests conveniently arranged for both simple or comprehensive automatic testing.

6-13 ENGINEERING EVALUATION (SEMIAUTOMATIC) TESTING

The two forms of testing examined in the previous sections constitute two extreme testing techniques: the *manual form*, where the test for each parameter can be set up separately, suitable for student laboratory work on individual ICs, and the automatic form, where quantities of ICs are to be tested for production or quality-control purposes in an automatic sequence type of operation, and calling for facilities for the requisite complex preprogramming of the varied tests provided by data cartridges or, in some cases, by a computer. For the in-between cases where a modest amount of ICs is to be evaluated by the user, whether in school or industrial laboratories, a more suitable form is a third kind of *semiautomatic testing*.

The operational amplifier tester illustrated in Fig. 6.15 (*Teledyne/Philbrick, model 5102*) is of the semiautomatic type, designed for conveniently obtaining engineering evaluation of the dynamic and the static characteristics of OP-AMP ICs in both monolithic and hybrid types. There are just five selector

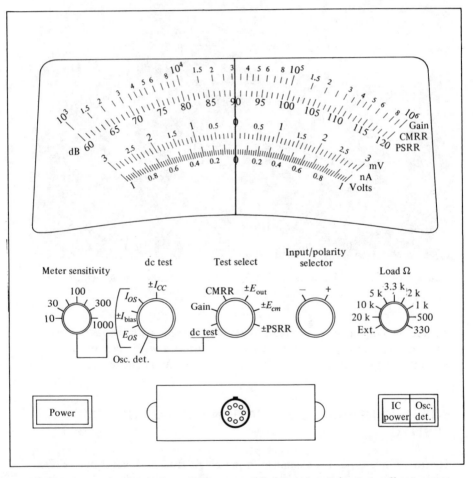

Figure 6.15 Operational-amplifier tester (semiautomatic): tests all OP AMPs (discrete and hybrid) for engineering evaluation, with meter readout for direct magnitude. (*Teledyne/Philbrick, model 5102*)

switches here to be set, after the proper amplifier socket has been plugged in. For testing any given parameter, the setting of three of the controls determines the automatic functioning of the proper circuit conditions for that parameter, while the operator simply sets the other two controls (METER SENSITIVITY and LOAD switches) for the evaluation at the desired operating conditions, as indicated in the tests given next.

DC Static Tests (As Set by DC TEST Switch)

The tests for *offset errors* (V_{io}, I_{io}, and I_B) are performed with the required zeroing for output done automatically. Tests for *current drain* ($\pm I_{CC}$) are measured under no-signal condition, giving a direct reading of current from either the positive or negative supply as selected by the *input polarity switch*. A check for *oscillation* (Osc Det) triggers a red light to indicate any IC oscillation that exceeds a 10 mV peak, 100 Hz to 10 MHz, with the amplifier connected for unity gain (100 percent feedback).

Dynamic Tests

The following parameters are determined dynamically under ac conditions by the use of low-frequency square waves and synchronous detection, with indications as noted for each setting of the *test select switch:*

 Gain (A_{VCL}) at full output, under load, with readout in decibels and volts per volt.

 CMRR is displayed both logarithmically and in decibels.

 Common-mode limits ($\pm E_{cm}$) are determined automatically for the voltage limits.

 Peak output voltage swing ($\pm v_o$) is determined automatically and read directly on the meter for either polarity.

 Power-supply rejection ratio (PSRR) is determined automatically with direct readout in decibels.

 Together with the proper test socket *accessories*, the tester thus provides a combination of manual and automatic operation for convenient evaluation of the linear IC device.

6-14 QUICK-CHECK TEST METHOD FOR OP AMPS [7]

For the purpose of checking widely used OP AMPs (of the general-purpose variety), the circuit given in Fig. 6.16 offers a simple, yet practical method that

[7] S. D. Prensky ' "Quick-Check" Test for OP AMPs with Your Multimeter,' *Radio-Electronics*, Sept. 1974; "Getting Started with OP AMPs," *Popular Electronics*, June 1975.

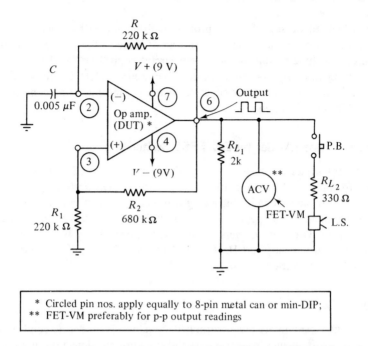

* Circled pin nos. apply equally to 8-pin metal can or min-DIP;
** FET-VM preferably for p-p output readings

Figure 6.16 Circuit diagram of quick-check for functional operation of many standardized types of OP AMP device under test (DUT).

rapidly verifies its functioning ability. In view of the complicated circuitry and the detailed operation involved in performing complete tests on LICs, this checking method gives a quick answer for determining confidence in the operating condition of a doubtful OP AMP and, as such, can well serve as a rough but practical *GO NO-GO check.*

The checking circuit employs the OP AMP as a multivibrator, generating a square-wave output. The simplicity of the system stems from the fact that the circuit uses only five active leads, out of the industry-standard eight pins in the popular forms of OP AMPs. Thus, the connections shown will apply to dozens of OP AMPs, where the eight-pin package (whether metal can or Mini-DIP) and the terminal numbering have become standardized by industry for these five active leads, as follows:

Inputs (inverting and noninverting) = ② and ③ ;

Output = ⑥, and ± V supply = ⑦ and ④, respectively.

Aside from these connections and the *RC* components, the circuit requires only a multimeter (preferably of the FET variety) and a small (2-in.) loudspeaker as accessories. With the push-button (PB) switch open, the meter reads the p-p value of the square-wave output in a nominally loaded condition. As a rule of thumb, this (p-p) value should read at least two-thirds of the supply voltage (or at least 12 V (p-p) from a ±9-V supply). Pressing the push-button switch calls for the OP AMP to furnish a reasonable flow of output current to produce an audible tone from the loudspeaker (around 1 kHz in this case); the output under this speaker-loaded condition should not drop to less than one half of the previous nominally loaded value [6 V (p-p) for this 18-V supply].

These two (p-p) meter readings, plus the audible tone, serve to establish the basic ability of the OP AMP to produce reasonable values of voltage and current output, and are thus able to verify the operating condition of the OP AMP in a simple manner.

As an added convenience, the *connections for the five active leads remain the same*, when checking practically all general-purpose types of OP AMPs, in spite of the various type numbers encountered from different manufacturers; this fortunate condition is due to the beneficial industry standardization of these popular devices. (Further details of construction and interpretation of test results are given in the article references at the beginning of this section.)

QUESTIONS

Note: For *design values* of typical OP-AMP characteristics in the equivalent model of Fig. 6.1, when using general-purpose OP AMPS (such as types 741, 748, and 101A), refer to Table 4-1.

6-1. A typical value for open-loop voltage gain is
   (a) >50,000
   (b) >10,000
   (c) >25,000

6-2. A typical value for input offset voltage is
   (a) <5 mV
   (b) <10 mV
   (c) <50 mV

6-3. A typical value of intrinsic input resistance is
   (a) >10 MΩ
   (b) >1 MΩ
   (c) >500 kΩ

6-4. Assuming a typical open-loop gain of 100,000, the value of loop gain will depend primarily on
   (a) Input resistance (R_{in})
   (b) Closed-loop gain (A_{VCL})
   (c) Feedback resistance (R_f)

6-5. In the test for open-loop gain in Fig. 6.2, the function of the attenuator is to
(a) Prevent saturation
(b) Prevent oscillation
(c) Allow a sufficiently large output voltage

6-6. In the test for closed-loop gain the values of the internal resistance of the signal generator
(a) Must always be kept low
(b) Can always be ignored
(c) Must be added as part of input resistance

6-7. In the quick-check method for testing OP AMPs, the device under test acts as
(a) An oscillator
(b) An inverting amplifier
(c) A noninverting amplifier

PROBLEMS

6-1. In the equivalent model of the intrinsic properties of an OP AMP (Fig. 6.1), assume a dc output of 10 V with no load ($R_L = \infty$).
(a) If the dc output drops to 5 V with a load resistor ($R_L = 600\ \Omega$, find the value of the equivalent output resistance (R_o).
(b) What value of dc output voltage would be expected theoretically if a load resistor (R_L) = 150 Ω were used?

6-2. In designing the closed-loop circuit of Fig. 6.3, state how a variable resistor for R_L can be used to find the output resistance (R_o) of the circuit.

6-3. In the closed-loop circuit of Fig. 6.3, the following values for the output (v_o) are obtained when only the load resistor (R_L) is varied; find the value of output resistance (R_o):
(a) When $R_L = 20\ k\Omega$, $v_o = 6$ V; when $R_L = 500\ \Omega$, $v_o = 3$ V
(b) When $R_L = 10\ k\Omega$, $v_o = 6$ V; when $R_L = 125\ \Omega$, $v_o = 1.2$ V

6-4. In Problem 6-3, what approximate value of input voltage (e_{in}) would produce the various output voltages (v_o) as stated?

6-5. In the circuit for measuring intrinsic R_{in}, Fig. 6.4, a value of 30 mV is used for the signal voltage (e_s). Find the value of this R_{in}, when R is adjusted as follows:
(a) $R = 500\ k\Omega$ and the terminal voltage (e_{in}) = 15 mV
(b) $R = 600\ k\Omega$, and the terminal voltage (e_{in}) = 10 mV

6-6. The quick-check circuit for testing OP AMPS, Fig. 6.16, acts as an oscillator because of positive feedback. Find what percentage of the output is fed back in the positive sense.

6-7. Find the theoretical frequency of the square-wave oscillation in Fig. 6.16, using the approximate expression

$$f_o = \frac{1}{2\pi RC}$$

Note: The actual operating frequency is also affected by the amount of positive feedback and the loading of the output in the circuit.

Power Amplifiers: Direct-Current and Audio

7-1 POWER LIMITATIONS IN INTEGRATED-CIRCUIT OPERATIONAL AMPLIFIERS

The previous discussions of general-purpose IC OP AMPs emphasized their most general use as *voltage amplifiers;* as such, the OP AMP ordinarily is not called upon to deliver any substantial amount of current, thus limiting its output power capability to a great extent. As an example of this limitation, we may estimate the power output obtained in the usual voltage-amplifier operation of a general-purpose OP AMP (such as the 741 type). With the customary dual power supply (±15 V), and with a recommended load resistor of 2 kΩ, we can expect a nominal voltage output of ±10 V (or 20 V peak to peak); this indicates an output current of only ±5 mA (or 10 mA peak to peak). Converted to rms values, this represents approximately 7 V at 3.5 mA, or *about 25-mW power output for the typical* OP AMP.

Of greater concern than this small value of output power is the *limited capability for substantial output current.* This is important in some dc applications, especially so in audio applications, where it is desired to avoid the use of an output transformer by feeding the current-operated voice coil of a speaker directly.

In cases where only a relatively small increase in output current is desired, there are two fairly simple options available in the OP-AMP type, without resorting to the power-amplifier type whose output is generally considered as exceeding 1 W. One method is the use of a *high-output-current monolithic type* of OP AMP or, alternatively, one can use the *power boosters* described next; however, for appreciable amounts of power (over 1 W) we must turn to linear ICs that are designated as *power amplifiers*, which are described later in this chapter.

7-2 HIGH OUTPUT-CURRENT OPERATIONAL AMPLIFIERS

Some OP AMPs offer a bit more in output-current capability than the typical OP AMP. For example, the *Analog Devices AD 517* offers a ±10-V output feeding a load resistor of 1,000 Ω, thus supplying ±10 mA, which is double the amount of current specified for this voltage output in the typical 741 type (the latter specifies a ±10-V output feeding loads equal to, or greater than, 2 kΩ). When feeding a lower-resistance load of 400 Ω, a maximum power output of up to 125 mW can be obtained. As a monolithic OP AMP, packaged in an eight-lead TO-99 (can)-type case, the *AD 517* would be quite suitable for certain limited-power dc or ac applications.

For a further increase of *output current up to as much as 1 A*, there is the *Fairchild µA791* OP AMP, which is available as a 10-lead device in a case similar to the familiar TO-3 case of power transistors to handle the large current output. Since it combines the input characteristics of the type 741 OP AMP as a preamplifier with a direct-coupled power amplifier in one package, it is discussed later as a power operational amplifier, useful for both dc and ac amplification. Similarly, the *CA3094/A* programmable power switch/amplifier provides high output current (to 300 mA peak to peak), and power output to 600 mW as a class A amplifier.

7-3 POWER BOOSTERS

Another method for increasing the current output, as a supplement to a circuit already using a general-purpose OP AMP, is that of adding a *power booster*. Here the use of Darlington emitter followers in a separate package serves to greatly increase the current gain of the overall amplifier; it provides unity voltage gains, thus retaining the voltage-gain characteristics of the general-purpose OP AMP.

One example of a convenient power booster is the *Burr-Brown 3329/03* type. (Although not a regular DIP package, it still fits the regular DIP socket.) When used to follow the OP AMP, it increases the output current capability of the combination to ±100 mA at ±10 V without the need for a heat sink. Since its characteristics allow it to be included within the feedback loop of an OP AMP (as shown in Fig. 7.1), it forms a very stable circuit for driving low-impedance loads. Its class B output stage ensures a minimum of quiescent power-supply

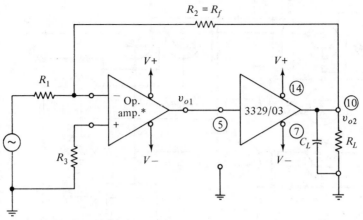

$$R_2 = R_f$$

Any general-purpose op. amp.

Figure 7.1 Power booster type (BB 3329/03): used with OP AMP for increasing output-current capability as in feeding a 50-ohm cable load (R_L). (*Burr–Brown*)

drain, while its low open-loop output impedance (10 Ω) enables it to drive a 50-Ω cable with its associated capacitance without degrading the frequency response of the OP AMP being used. Such an output, incidentally, is quite suitable for driving a small (45-Ω) loudspeaker to a satisfactory low-volume listening level.

A similar power-booster type is produced by *Motorola* (*MC1438R*) for output currents up to ±300 mA with a 1.5-MHz bandwidth (also at unity gain). This comes in a nine-pin package (looking like a small TO-3 power-transistor can), and requires a special nine-pin socket.

7.4 POWER OPERATIONAL AMPLIFIER

A device with an output-current capability of a full ampere is the *Fairchild μA791* OP AMP, packaged as a 10-lead device in a case similar to the familiar TO-3 case of power transistors. It combines a type 741 preamplifier and a power amplifier in the one package. As shown in the schematic diagram of Fig. 7.2, it is direct coupled throughout and can therefore be used as a servo amplifier at very low frequencies, for dc uses, and in audio-amplifer applications with a proper heat-sink, as discussed later. It is thermal and short circuit protected, having a current-sensing terminal allowing selection of the sensing resistor (R_{SC}) for limiting the amount of short-circuit current to 500 mA (when R_{SC} = 1.5 Ω) or to 1 A (with R_{SC} = 0.7 Ω). With R_{SC} equal to zero, the specified large-signal

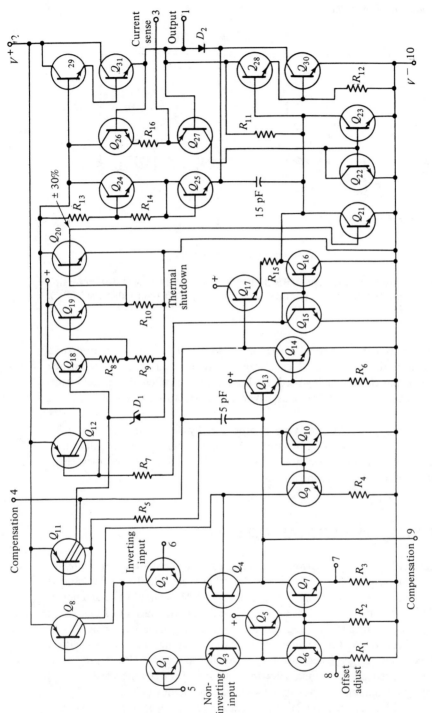

Figure 7.2 Equivalent circuit of power operational amplifier, μA791: it combines the input characteristics of a 741 type OP AMP as a preamplifier with an output current capability of ±1 A at ±11 V, while retaining its dc and ac characteristics as an OP AMP. (*Fairchild Camera and Instrument Corporation*)

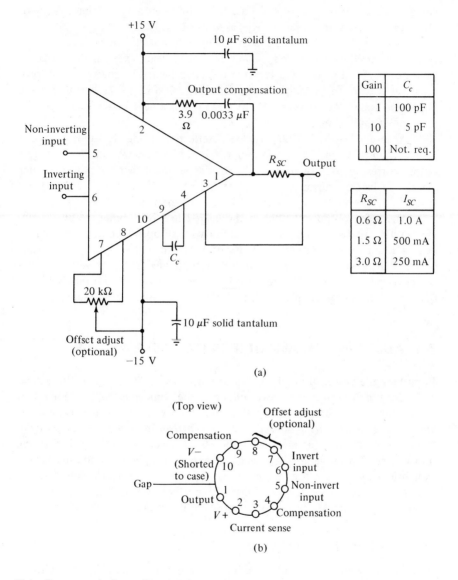

Figure 7.3 Connection diagram for μA791 power operational amplifier: (a) current-sensing resistor (R_{SC}) limits the short-circuit current and output compensation of 3,300 pF is in series with a 3.9-Ω resistor; (b) socket connections for 10-lead, TO-3 type package. (*Fairchild Camera and Instrument Corporation*)

voltage gain exceeds 20,000 V/V, and will provide an output voltage swing of over ±11 V into a load resistance (R_L) as low as 11 Ω.

The connection diagram, shown in Fig. 7.3, indicates that it has some internal compensation similar to type 741, so that the usual external compensating capacitor (C_c) can be omitted for gains of 100 or higher; however, an additional capacitor is required for *output compensation* (3,300 pF in series with 3.9 Ω). Moreover, again like the type 741, offset-null adjustment is provided when desired for dc applications.

Accordingly, the *μA791* may be described as being the equivalent of a combination of a type 741 preamplifier, followed by a power booster having an output-current capability of 1 A. (It should be noted that the power booster mentioned in the previous section, *Burr–Brown 3329/03*, can be added to any type of OP AMP and is not restricted to the 741 type.)

The input–output relations of the *μA791* may be summarized by describing the device as a direct-coupled amplifier with an open-loop input impedance of 2 MΩ and with an output capability of ±11 V into 11 Ω, or a power output of 5.5 W under maximum operation with heat sinking. This high-gain and high-power-output capability is obtained in this power OP AMP at a relatively low quiescent (zero-signal) supply current of 30 mA from a dual ±15-V supply.

7-5 AUDIO POWER AMPLIFIERS (TO 1 OR 2 WATTS)

In pushing the power of amplifiers up to outputs of 1 W or higher, the general difficulty lies in problems of heat buildup in the semiconductor junctions. This is particularly important in integrated circuits, because of their compactness, where the *generated heat is necessarily confined on a very small silicon chip*. Satisfactory methods for meeting this problem have been developed, resulting in a variety of package and heat-sink forms; presently, both monolithic as well as hybrid versions have been made available, going beyond the previous levels of 1 or 2 W and providing devices of 8-W and greater outputs. These devices are in a form compatible with the compact layouts of the standardized OP AMPs.

One- and Two-Watt Amplifiers

As a start in tracing the trend toward higher powers (and passing by the *MC1306*, $\frac{1}{2}$-W level), we can start with the early audio amplifiers at the 1-W power level. A typical example at this level is the *Motorola MC1554G/1454G*, which comes in a 10-lead, metal-can package. It is designed to amplify signals to 300 kHz, delivering 1 W to a typical load of 16 Ω, either direct or capacitively coupled. Gain options of 10, 18, or 36 V/V are provided at single-supply operation (V+ = 16 V) or, alternatively, at a split supply of ±8 V. For consumer audio systems in this 1-W class, there are also available plastic DIP versions; examples include the *Texas Instruments SN76011* (with heat-sink tab) for 1 W into an 8-Ω load, operating from a 4- to 13-V supply.

The 2-W power level is often used for relatively inexpensive audio systems, such as for a phono amplifier. A typical IC audio amplifier for this level is the *National LM380*, which aims at a low parts count for consumer applications when feeding an 8-Ω speaker. In the diagram for a phono amplifier application (Fig. 7.4), the connections for the 14-lead DIP package of the *LM380* are shown. In this case, the heat-sink requirement is met without the use of a separate tab; instead, three pins in the middle of each side of the standard DIP package are soldered together (pins 3, 4, 5 and 10, 11, 12). When mounted on epoxy-glass board, the maximum junction temperature of 125°C is controlled by soldering the six pins together to the copper foil of the mounting board (minimum surface of 6 in.²). The quiescent supply current for the *LM380* has been brought down to 7 mA. Another version of a 2-W amplifier in the *Sprague ULN-2280B*.

For stereo amplifiers there are dual versions, having both 2-W channels in a single 14-lead DIP package. The *National LM377* is an example of such a type, where each channel feeds an 8-Ω speaker, returned to ground from a supply voltage of from 9 to 30 V.

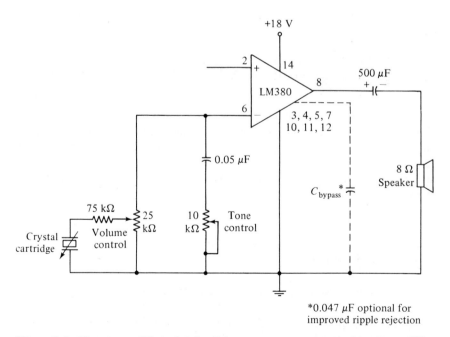

*0.047 μF optional for improved ripple rejection

Figure 7.4 Phono amplifier circuit of low component count: 2-watt amplifier (LM380), with loud speaker connection returned to ground; the middle pins on each side of the DIP package (pins 3, 4, 5 and 10, 11, 12) are soldered together for heat-sinking. (*National Semiconductor*)

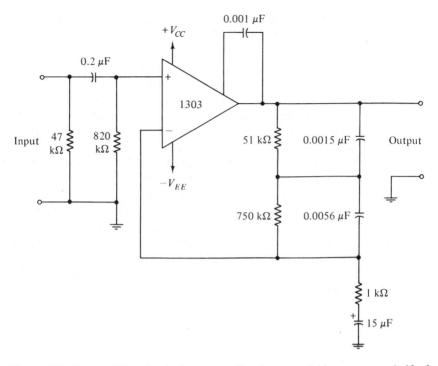

Figure 7.5 Preamplifier circuit for magnetic phono cartridge uses one half of dual stereo preamp MC1303, with RIAA equalization. (*Motorola Semiconductor Products Inc.*)

    To accompany dual audio amplifiers for stereo use, there are companion *dual preamplifiers*, containing both preamp channels in the one DIP package. Examples of such dual preamplifiers are the *Motorola MC1303* and the *National LM381* types, both designed for stereo use. A circuit diagram using half of the *MC1303* to preamplify a *magnetic phono cartridge* with RIAA equalization is shown in Fig. 7.5.[1]

7-6 HIGHER-POWER INTEGRATED-CIRCUIT AUDIO AMPLIFIERS

Monolithic versions of audio power amplifiers typically up to the 8-W level find use in such consumer applications as auto radios and TV receivers, and also in some industrial applications. While still in the favored DIP form, the IC package is necessarily modified to accommodate heat-sink considerations at this power

[1] B. R. Rogen, "Op Amps at Work: 10 Audio Circuits," *Radio Electronics*, Dec. 1972.

level. However, the developments in IC fabrication have succeeded in producing compact packages that are vastly more convenient than any of the discrete forms of amplifiers.

A typical 5-W IC audio amplifier, *Fairchild $\mu A706$*, is shown in schematic form in Fig. 7-6. It consists basically of a *preamplifier and power amplifier.* The preamp has high-gain transistor Q_2 acting as a common emitter with a high-impedance collector load provided by current source Q_3. Transistor Q_1 acts as an input-buffer transistor, thus providing a high input resistance (3 MΩ) referenced to ground.

The *power-amplifier* section feeds power transistors Q_{13} and Q_{14} as a *quasi-complementary push–pull amplifier stage.* (The term "quasi" is used to indicate that the circuit operates in complementary form, even though both power transistors are *NPN*, instead of the usual complementary combination of *NPN* and *PNP* types.)

The overall closed-loop voltage gain of the amplifier is determined initially by the internal negative feedback applied through resistor R_6 from the output to the emitter of Q_2. This internal feedback reduces the high open-loop gain by a factor of about 50 to a value of A_v = 200 or 46 dB (obtained from the ratio of R_6/R_3). A gain-control option is provided at terminal 8 for the external connection of a gain-reducing resistor (R_B in series with a 100-μF capacitor). Thus, as shown in the data box of Fig. 7.7 with R_B = 100 Ω, the gain is reduced from 46 to 34 dB (or from 200 to 50 V/V).

Operating from a single power supply (V+ = 14 V), the 5-W output can feed a 4-Ω speaker connected between output and supply, as in Fig. 7.7(a); alternatively, with the addition of a 47-Ω resistor and 220-μF capacitor, the speaker can be connected between output and ground as shown in part (b). In the latter grounded-speaker connection, an optional tone control is included. Under test-circuit conditions, a value of C_S equal to 5.6 nF (5,600 pF) provides for a 3-dB falloff in gain at 4 kHz.

Examining the purpose of the other external components in Fig. 7.7(a), there are the two filter (or ripple) capacitors of 100 μF each, the output coupling capacitor of 1,000 μF, and the boot-strapping capacitor of 0.33 μF. In addition, we have compensation capacitor C_c at the terminal 6 in conjunction with capacitor C_F at the same terminal. The test-circuit values for these (at the maximum gain of 46 dB) are given as C_c = 1.5 nF (or 1,500 pF) and C_F = 150 pF; corresponding values of C_c and C_F are given for reduced gain (A_v) or reduced bandwidth (BW).

Using the test-circuit values (46-dB gain and 20-kHz bandwidth at a supply voltage of 14 V), the device can deliver 5.5 W into the 4-Ω load, with less than 10 percent total harmonic distortion (THD). This high value of distortion is caused by clipping; however, it is radically reduced to a THD value of less than 2 percent at an output-power level of 4 W, and still further reduced to below 1 percent at power levels of 3 W or less.

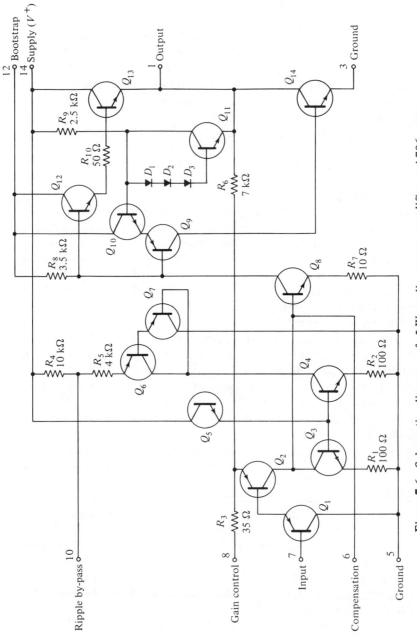

Figure 7.6 Schematic diagram of 5-W audio power amplifier, μA706; constructed on a single silicon chip, it provides 5.5-W maximum power output into a 4-Ω speaker (see text for heat-sinking options). (*Fairchild Camera and Instrument Corporation*)

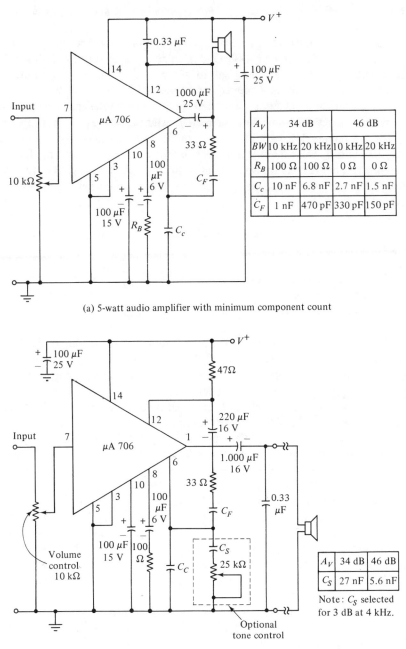

(a) 5-watt audio amplifier with minimum component count

A_V	34 dB		46 dB	
BW	10 kHz	20 kHz	10 kHz	20 kHz
R_B	100 Ω	100 Ω	0 Ω	0 Ω
C_c	10 nF	6.8 nF	2.7 nF	1.5 nF
C_F	1 nF	470 pF	330 pF	150 pF

A_V	34 dB	46 dB
C_S	27 nF	5.6 nF

Note: C_S selected for 3 dB at 4 kHz.

(b) 5-watt audio amplifier, with 4 Ω load connected to ground

Figure 7.7 Connections for μA706 audio power amplifier (5.5 W max): (a) with loudspeaker connected to V+ for minimum component count; (b) with loudspeaker connected to ground (see text for explanation of capacitor symbols). (*Fairchild Camera and Instrument Corporation*)

151

As a device designed for consumer applications, the package has been kept compact in the familiar 14-lead dual-in-line plastic form, but this convenient DIP form has been modified to accommodate various heat-sinks. In the basic (package A) form, a copper slug is exposed at the top of the package, providing a thermal conduction path from the junction to the case. To improve the thermal junction-to-ambient thermal resistance (Θ_{J-A}) of package A (73°C/W), a copper bracket soldered to the topside is optionally offered as package B, reducing this thermal resistance to 55°C/W. Various heat-sink devices are commercially available for attachment to the bracket of package B to further reduce the thermal resistance, as required by specific applications, to satisfy the relation

$$\Theta_{J-A} \leqslant \frac{T_d - T_A}{P_d},$$

where T_d, maximum junction temperature = 150°C,
T_A = ambient temperature,
P_d, maximum device dissipation = 3 W.

For ambient temperatures reasonably below the maximum of 85°C, a satisfactory thermal resistance of 22°C/W can be obtained by soldering the bracket of package B to the copper side of a two-sided printed-circuit board. Moreover, at power-output levels smaller than the maximum of 5.5 W, satisfactory operation of either package A or B can be obtained in particular situations without employing additional heat sinking.[2]

Similar audio power amplifiers up to 4 or 5 W, but which still retain the convenient package form of monolithic LICs, are exemplified by such devices as the *Texas Instruments SN76024* (4 W). An example of a dual audio amplifier is the *LM378* (4 W/channel) for use in stereo amplifiers.

An example of a monolithic IC 8-W power amplifier is the *Motorola TDA2002* device; a block diagram of the unit is provided in Fig. 7.8. Operating in class B, the TDA2002 is intended for automotive and general-purpose audio applications. A current capability of 3.5 A permits the device to drive a low-impedance load (down to 1.6 Ω) with low harmonic and crossover distortion. The TDA2002 is available in a plastic package attached to a heat-sinking tab (Fig. 7.9).

The device provides an output power of 8 W with R_L = 2 Ω and a minimum of 4.8 W with R_L = 4 Ω, at 14.4 V. In addition to operating over a supply voltage range of 8 to 18 V, the TDA2002 features internal overload protection, short-circuit current limiting, and supply overvoltage protection.

[2] Fairchild Application Note 317 illustrates commercially feasible layouts for both a single 5-W μA706 amplifier and also for two devices in an FM stereo receiver.

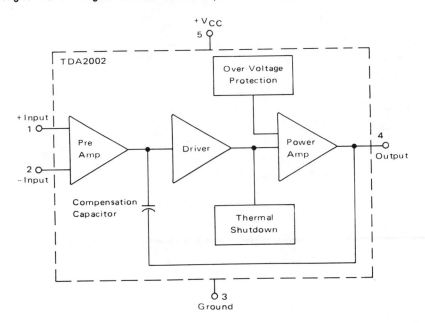

Figure 7.8 Block diagram of the TDA2002 monolithic 8-W audio power amplifier. (*Motorola Semiconductor Products, Inc.*)

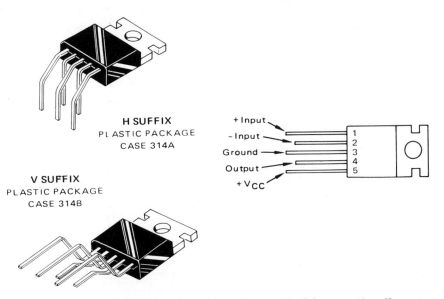

Figure 7.9 TDA2002 8-W amplifier: (a) package styles; (b) connection diagram. (*Motorola Semiconductor Products, Inc.*)

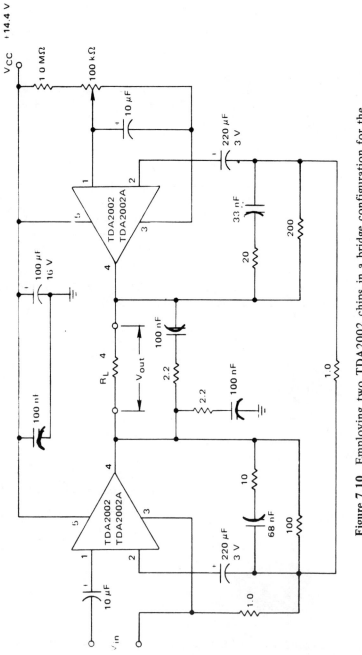

Figure 7.10 Employing two TDA2002 chips in a bridge configuration for the realization of a 15-W audio amplifier. (*Motorola Semiconductor Products, Inc.*)

By using two TDA chips in a bridge circuit (Fig. 7.10), a 15-W power amplifier is realized where the output is taken across a 4-Ω load resistance. For IC amplifiers designed to deliver more than 15 W, thick-film technology is generally used.

QUESTIONS

7-1. The maximum sustained output current that can usually be expected from general-purpose OP AMPs (such as the 741) is
  (a) Around ±5 mA
  (b) Around ±25 mA
  (c) Around ±100 mA

7-2. The recommended load resistor for full nominal ±10V output for the 741 type OP AMP is
  (a) 10 kΩ
  (b) 5 kΩ
  (c) 2 kΩ

7-3. For OP AMPs with lower bias current (I_B) or voltage offset (V_{io}) (such as types 101A and 108), the recommended load resistor for nominal ±10-V output is
  (a) 10 kΩ
  (b) 5 kΩ
  (c) 2 kΩ

7-4. The nominal power output for general-purpose OP-AMPs is
  (a) 10 mW
  (b) 25 mW
  (c) 100 mW

7-5. The output impedance (R_o) of a power-booster type of IC (such as a Burr-Brown 3329/03) is around
  (a) 10 Ω
  (b) 100 Ω
  (c) 500 Ω

7-6. A direct-coupled power amplifier (such as the 791 type) can be used to amplify
  (a) dc signals only
  (b) ac signals only
  (c) Both dc and ac signals

7-7. Audio power-amplifiers rated for 5 W are found
  (a) In both monolithic and hybrid types
  (b) Only in monolithic types
  (c) Only in hybrid types

PROBLEMS

7-1. In a power-amplifier OP AMP circuit, an output of 12 V(p-p) is obtained with a load resistor of 400 Ω. When a load resistor of 8 Ω is substituted, the output voltage is found to drop to 6 V(p-p). Find the approximate output resistance (R_o) of the amplifier. (*Hint:* At the relatively high load-resistance value of 400 Ω, consider the voltage drop across output resistance R_o as negligible.)

7-2. In the circuit diagram of a type 706 (5-W) audio amplifier (Fig. 7.6), identify the drive and output transistors which account for the output designation of "quasi-complementary" circuit.

7-3. In the circuit of the type 380 phono amplifier (Fig. 7.4), a signal of 250 mV ac is fed into pin 6. If the output voltage across the 8-Ω speaker is 4 V ac (considering the impedance of the 500-μF capacitor as negligible), find
   (a) The power output (P_o)
   (b) The peak-to-peak voltage to be expected from an oscilloscope across the speaker
   (c) The voltage gain of the circuit

7-4. In the circuit of Fig. 7.4, using the stated values of +18 V for the power supply and 7.5-mA quiescent circuit, find the no-signal power dissipation (P_D).

7-5. In the 380 circuit (Fig. 7.4), state the purpose of pins 3, 4, 5, 10, 11, and 12, and their relation to the heat-sinking requirements for a full 2-watt output.

Chapter 8

Consumer Communication Circuits

8-1 IMPACT OF LINEAR INTEGRATED CIRCUITS ON COMMUNICATION EQUIPMENT

Owing to the continuing advances made in monolithic LIC technology, the realization of, for example, an AM receiver on a silicon chip is a reality. These advances have resulted in new and improved circuits for consumer and communications applications. In addition to AM, LICs have been developed for FM, TV, and stereo applications. These circuits provide improved electrical performance, high reliability, and low cost. Troubleshooting also has been greatly simplified. Locating a defective component in a discrete section of, for example, a television receiver is now reduced to replacing a defective LIC with a new one.

Acceptance of LICs was a much slower process in communication equipment than in the case of digital computer ICs. There is far less standardization in communication designs and a more diverse group of equipment manufacturers is involved than in the case of computers. Another early roadblock was the inability of LICs to effectively perform frequency-selective filtering internally. While digital computer circuits require fast switching speeds, they are simply off-on switches. The LIC in communication systems, however, must compete with individual discrete components specifically optimized for each part of the circuit. For example, a radio receiver may use a low-noise high-frequency RF amplifier transistor in its "front end" that has been optimized for resistance to cross-modulation signals (interference in which the desired carrier signal is modulated by an undesired signal having a different carrier frequency). This transistor, however, would not make a good audio output amplifier. The same discrete component receiver might use a low-frequency, high-current pair of audio output transistors, made with a process radically different from that of the RF amplifier, to drive the loudspeaker. Monolithic circuits, however, are constrained to a common process. Thus communication LICs are often compromised in performance, compared to their specially designed discrete counterparts.

A further problem, especially in television receivers, is the need to handle high power and high voltages in circuitry interfacing with the cathode-ray tube. While LICs have been developed to approach the power and voltage capability of discrete circuits, practical television receivers still use discrete components in such functions as a matter of economics. While modern LIC techniques permit certain frequency-selective functions to be performed monolithically, economics have continued to favor the use of conventional external tuning elements where needed. There is a trend for such tuning to be lumped into single, highly selective blocks when connected to the internally complex LIC, as opposed to the conventional "interstage tuning" of discrete circuitry, in which a number of "cans" are interspersed between discrete transistor stages. (Consolidation of tuning elements in communication LICs reduces the number of IC terminals needed to make connection between internal and external elements to a minimum.)

Integrated-Circuit Subsystems

Monolithic communication LICs are, however, coming into their own, as more and more separate functions are combined on a single IC chip. Drawing a parallel to early computer ICs, the replacement of a single function (whether it is a digital gate or a single IF amplifier stage in a radio receiver) was not any less expensive than performing the task with discrete transistors and associated components. Just as computer circuits became more complex, with two, then four, and now thousands of gates in a package (as in *very large scale integration*, or *VLSI*), LIC communication circuits are replacing discrete components by building complete *subsystems* on a chip. One example of a "radio receiver" LIC, the *Ferranti ZN414*, is a 10-transistor TRF receiver in a three-lead TO-18 package (Fig. 8.1). While the manufacturer suggests use with an audio amplifier, the IC can drive a sensitive earphone, as illustrated in Fig. 8.2.

Another example is a single-chip AM/FM LIC developed by the *General Electric Company*. Some 60 transistors in a dual-in-line plastic package provide the AM mixer, AM local oscillator, AM/FM IF amplifier, AM/FM detector, and the audio amplifier functions. The only active devices external to the IC are located in the FM VHF tuner, where two low-cost transistors are used. A block diagram of an AM/FM radio using the LIC is provided in Fig. 8.3.

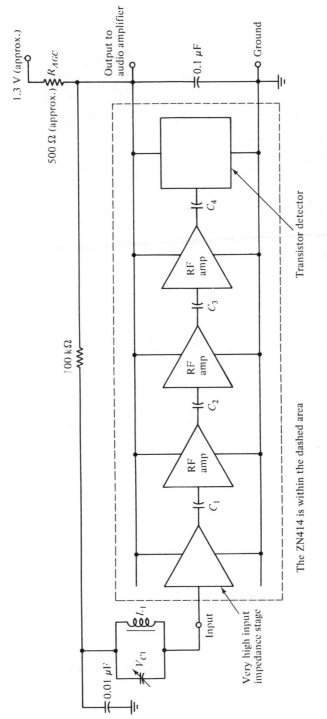

Figure 8.1 Functional diagram of the ZN414, a 10-transistor tuned radio frequency (TRF) linear integrated circuit which is available in a three-pin TO-18 transistor package. (*Ferranti Limited*)

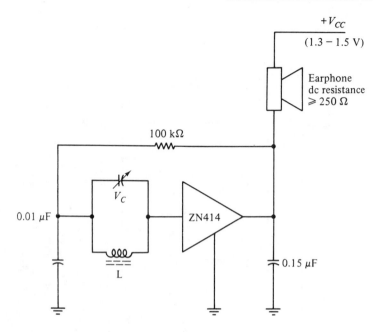

Figure 8.2   Simple earphone radio using the ZN414 chip. (*Ferranti Limited*)

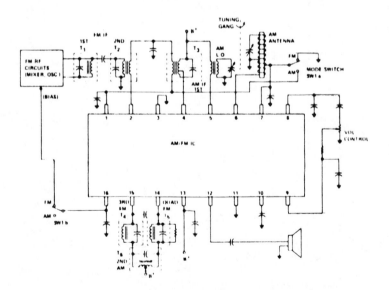

Figure 8.3   An AM/FM receiver using an LIC containing some 60 transistors. (*General Electric Company*)

8-2 BROADCAST FM RECEIVER CIRCUITS

RF/IF Amplifiers

The earliest LIC to receive mass usage in consumer electronic equipment was the type *μA703* (Fig. 8.4), which is a simple differential, or emitter-coupled, amplifier intended to replace a single transistor or tube in the 10.7-MHz "IF

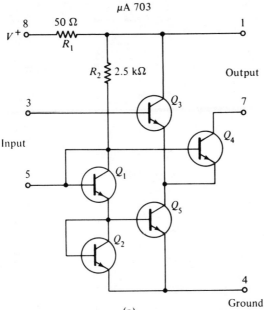

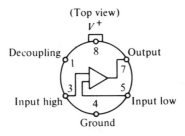

Figure 8.4 RF/IF amplifier (μA703): (a) schematic diagram of monolithic amplifier, usable in limiting or nonlimiting applications to 150 MHz; (b) connection diagram. (*Fairchild Camera and Instrument Corporation*)

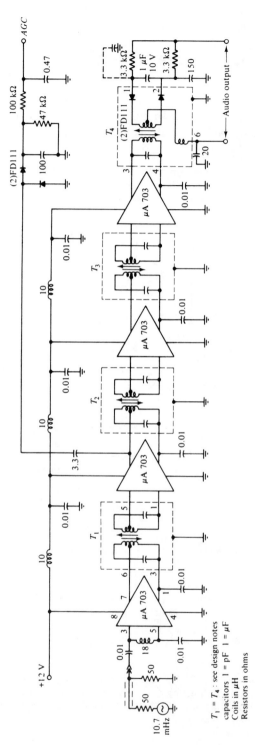

Figure 8.5 Four-stage FM-IF amplifier using IC type 703, showing automatic gain control (AGC) and FM detector connections. (*Fairchild Application Bulletin APP-151*)

strip" of a high-fidelity FM receiver. Typical of a class of devices known as RF/IF amplifiers, the μA703 requires conventional LC tuned "IF cans" for interstage coupling and to complete its dc biasing. Transistors Q_3 and Q_4 constitute the differential amplifier stage; Q_5 is a constant current source which is stabilized by transistors Q_1 and Q_2 connected as diodes.

Although the type 703 was not cost competitive with discrete transistor stages at the time of its introduction, it first found its way into higher-priced FM receivers, and later became universally used because of its superior FM limiter action. A conventional single-transistor FM-IF amplifier is required to produce a constant-amplitude "limited" output, even when the amplitude of the input signal varies. It does so by switching rapidly from cutoff (no conduction) to saturation (maximum conduction) when driven by small input signals. Saturation, unfortunately, is a mechanism that can vary with the strength of the incoming signal, since a transistor comes out of saturation slowly because of stored charge in its base. It thus presents a varying impedance to the tuned circuit which is driving the input, causing variations in tuned-circuit frequency and bandwidth with varying input-signal strength. The type 703, however, is a differential amplifier biased so that it is not permitted to saturate, but merely shunts the current from the constant-current source into either the input or output transistor of the pair. Thus excellent limiting action is obtained, compared to discrete transistor amplifiers, without any of the built-in discrete amplifier problems. A typical application is shown in Fig. 8.5.

8-3 FM-IF STRIPS

Newer FM broadcast receivers use more complex "IF strip" LICs (such as the *RCA CA3043* shown in Fig. 8.6), containing several direct-coupled differential-amplifier limiting stages, a common dc biasing source, and often FM detectors and audio preamplifiers. There is essentially no difference between the FM broadcast IF at 10.7 MHz and the broadcast-television sound IF at 4.5 MHz, and the same types of IF strips are used interchangeably in both FM broadcast and television receivers. Many others have essentially the same structure, but provide different gains and various types of FM detectors. Many designers of stereo FM tuners use the amplifier portions of such strips, adding their own specialized discrete-component detector circuits. Certain receivers may follow the IF strip with a phase-locked-loop (PLL) type of FM detector, discussed later in the chapter.

Practical application of these high-gain FM-IF strips requires very careful location of components and interconnections, as well as strategic shielding. While these strips have about as much voltage gain as an IC OP AMP (80 dB or more), the OP AMP has rather low frequency response, while the IF strip extends to the tens of megahertz, where capacitive and inductive coupling between input

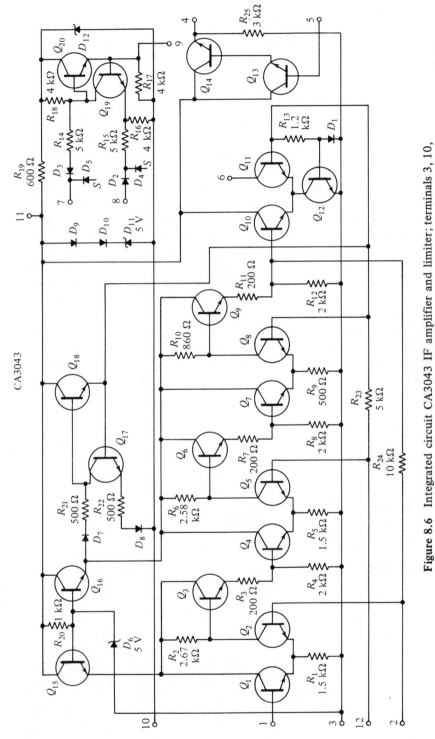

Figure 8.6 Integrated circuit CA3043 IF amplifier and limiter; terminals 3, 10, and substrate should be connected to the most negative point. (*RCA Solid State Division*)

164

and output is a significant problem. Most IF strips require only a few microvolts of input to produce volts of output. Thus there is a problem in making operating measurements with an oscilloscope probe, as enough pickup is added to give misleading results. Input and output are separated by a fraction of a centimeter on the IC, as opposed to several centimeters in a conventional "IF can" type of strip.

Since the IC form of the IF strip has several direct-coupled gain stages, it requires a "lumped" external bandpass filter, as in Fig. 8.7. The filter is usually a multisection ceramic or crystal filter, giving very sharply shaped band-pass characteristics and relatively flat phase shift versus frequency response. Such characteristics are usually dependent upon the driving and load impedances presented to the filter. Input resistance and capacitance of the monolithic IF strip, therefore, must be controlled to give proper filter response.

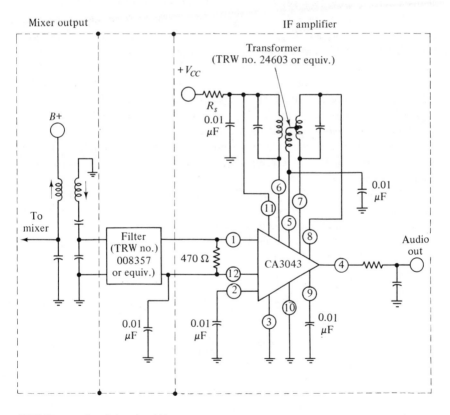

NOTE: Pin nos. apply to 8-pin package (either 8-pin metal can or 8-pin mini-DIP types).

Figure 8.7 FM receiver design, using conventional tuner, an LC filter, and a CA3043 as IF amplifier and limiter, feeding a discriminator transformer. (*RCA Electronic Components Publication No. ST-3773*)

Because the strip is usually direct coupled and has very large gain, extremely small dc input offsets would be sufficient to drive the output to either its positive or negative limit, where it no longer amplifies signals. To avoid this problem, dc feedback is used, in which a negative-feedback resistor from output to input automatically adjusts the effective input voltage to compensate for any existing input offset conditions, thus guaranteeing that each stage will always operate in the center of its symmetrical limiting characteristic. Such feedback is desirable only for direct current; ac feedback would reduce ac voltage gain, which should be as high as possible. Thus a decoupling capacitor is used in the feedback loop to short out any 10.7-MHz output that might otherwise travel back to the input, thus leaving only the dc component. The capacitor used for this purpose must be of high quality at 10.7 MHz, and should be connected with very short leads. Otherwise, enough IF output will be fed back to degrade performance or, possibly, to allow the high-gain strip to oscillate.

An example of an LIC which, in a single chip, includes an IF amplifier, quadrature detector, audio preamplifier, and circuits for AGC, AFC, tuning meter, deviation-noise muting, and ON channel detection is the *RCA CA3189E*. Referring to the block diagram of Fig. 8.8, the chip includes a three-stage FM-IF amplifier-limiter configuration with a level detector for each stage, a doubly balanced quadrature FM detector, and an audio preamplifier that features the optional use of a muting (squelch) circuit. In addition, the 3189E features a programmable delayed AGC for the RF tuner, an AFC drive circuit, and an output signal to drive a tuning meter and/or provide stereo switching logic. The use of the 3189E in an FM-IF receiver section and its printed-circuit board layout is illustrated in Fig. 8.9.

An example of a narrow-band FM-IF strip designed for use in voice communication scanning dual-conversion receivers is the *Motorola MC3357* (Fig. 8.10). The mixer-oscillator circuit converts the input frequency of 10.7 MHz

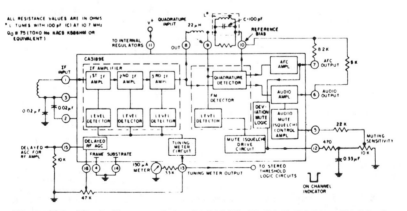

Figure 8.8   Block diagram of the CA3189E. (*RCA Solid State Division*)

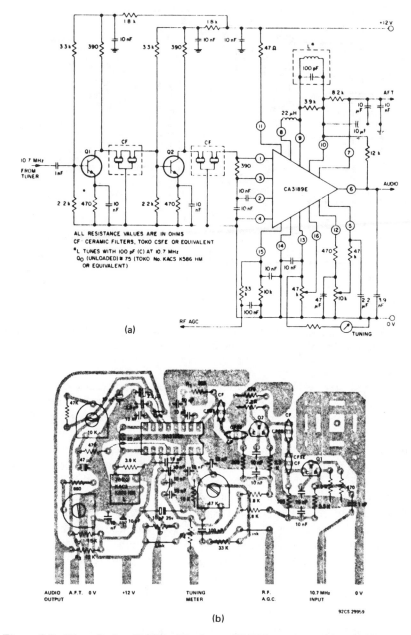

(a)

(b)

92CS 29959

Figure 8.9 Use of the 3189E chip in an FM-IF receiver section: (a) circuit; (b) PC board layout. (*RCA Solid State Division*)

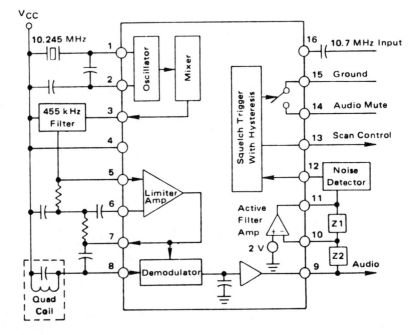

Figure 8.10 Functional block diagram of the MC3357 narrow-band FM-IF chip. (*Motorola Semiconductor Products, Inc.*)

down to 455 kHz, where, after external band-pass filtering, most of the amplification is done. The audio is recovered using a conventional quadrature FM detector. The absence of an input signal is indicated by the presence of noise above the desired audio frequencies. This *noise band* is monitored by an active filter and detector. A squelch trigger circuit indicates the presence of noise, or a tone, by an output which can be used to control scanning. At the same time, an internal switch is operated which can be used to mute the audio.

The oscillator is an internally biased Colpitts type with the collector, base, and emitter connections at pins 4, 1, and 2, respectively. A crystal can be used in place of the usual coil. The mixer is doubly balanced to reduce spurious responses. The input impedance at pin 16 is set by a 3-kΩ internal biasing resistor and has low capacitance, allowing the circuit to be preceded by a crystal filter.

8-4 FM STEREO MULTIPLEX DECODERS

The FM stereo decoder, or demodulator, is an almost ideal candidate for monolithic fabrication; it must be relatively complex and it demands accurate matching of components for best performance. An example is the *Motorola MC1310*

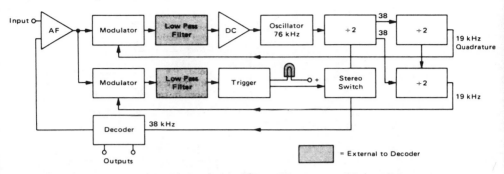

Figure 8.11 Block diagram of the MC1310 FM stereo demodulator. (*Motorola Semiconductor Products, Inc.*)

FM stereo demodulator. Referring to the block diagram of Fig. 8.11, the oscillator is running at 76 kHz and, feeding through two divide-by-2 stages, returns a 19-kHz signal to the input modulator. There the returned signal is multiplied with the incoming signal so that when a 19-kHz pilot tone is received, a dc component is produced. The dc component is extracted by the low-pass filter and used to control the frequency of the internal oscillator, which then becomes phase locked to the pilot tone. With the oscillator locked to the pilot, the 38-kHz output from the first divider is in the correct phase for decoding a stereo signal. The decoder is essentially another modulator in which the incoming signal is multiplied by the regenerated 38-kHz signal. The regenerated 38-kHz signal is fed to the stereo decoder via an internal switch, which closes when a sufficiently large 19-kHz pilot tone is received.

The 19-kHz signal returned to the 38-kHz regeneration loop modulator is in quadrature with the 19-kHz pilot tone when the loop is locked. With the third

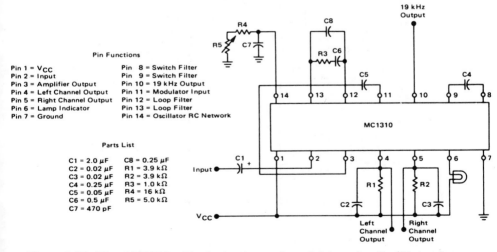

Figure 8.12 The MC1310 chip in a stereo demodulator circuit. (*Motorola Semiconductor Products, Inc.*)

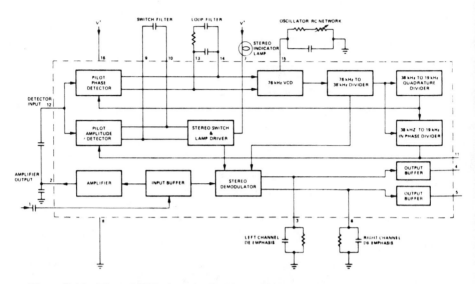

Figure 8.13 The μA758 phase-locked-loop FM stereo multiplex decoder. (*Fairchild Camera and Instrument Corporation*)

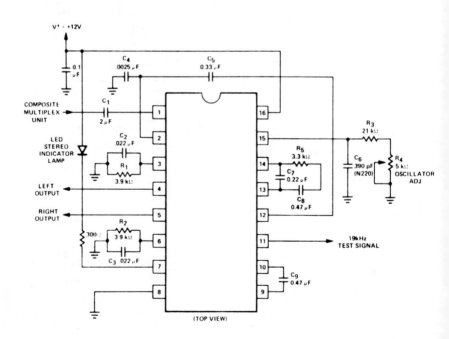

Figure 8.14 Application of the μA758 chip in a decoder circuit. (*Fairchild Camera and Instrument Corporation*)

divider stage appropriately connected, a 19-kHz signal in phase with the pilot tone is generated. This is multiplied with the incoming signal in the stereo switch modulator, yielding a dc component that is proportional to the pilot tone amplitude. This component, after filtering, is applied to the trigger circuit, which activates both the stereo switch and an indicator lamp. An example of its application is provided in Fig. 8.12.

A block diagram of a phase-locked-loop FM stereo multiplex decoder (*Fairchild μA 758*) is illustrated in Fig. 8.13. Among its features, the chip provides 45-dB channel separation and automatic stereo-mono switching. The device decodes an FM stereo multiplex signal into right and left channels while suppressing SCA information which may be present in the composite input signal. Typical application of the LIC is shown in Fig. 8.14.

8-5 BROADCAST AM RECEIVER CIRCUITS

Unlike the FM-IF strip, the IF function in an AM receiver does not require as large a gain, nor does it require limiting to remove the AM component of the signal. Instead, there is a requirement to linearly amplify the AM signal over a very wide variation in signal strength while maintaining a constant output. This is accomplished by use of automatic gain control (AGC). In this action, a dc voltage, proportional to the amount of signal reaching the detector, is returned to one or more IF amplifiers, whose gain can be controlled by the dc voltage. Thus, excessively strong signals reduce the IF gain; as a result, gain is automatically adjusted over a wide range of input signals.

Automatic Gain Control in RF/IF Amplifiers

In discrete AGC amplifiers, gain is reduced by either increasing transistor current above the optimum gain point or by reducing transistor current to very low values. In both cases, the input impedance and signal-handling characteristics of the single transistor stage vary considerably as gain varies, upsetting the tuning characteristics of the receiver. A monolithic LIC designed to overcome these difficulties is the RF-IF amplifier with AGC, typified by the *MC1550* (Fig. 8.15). It is a differential amplifier; however, the current source (Q_1) is the high-frequency input amplifier, while one of the differential pair (Q_2) is a common-base amplifier connected to the common-emitter current source, thus forming a transistor analogy of the well-known "cascóde" circuit. With negligible AGC voltage applied to its base (terminal 5), the transistor not connected to the output (Q_3) is biased fully off, and the circuit achieves maximum gain. With increased AGC voltage, the differential amplifier is balanced, and half the dc current from the current source is diverted into the nonoutput side (terminal 9), as well as half the output signal (a gain reduction of 2, or −6 dB). As the nonoutput side is increased, more and more signal current is routed away from the

MC1550

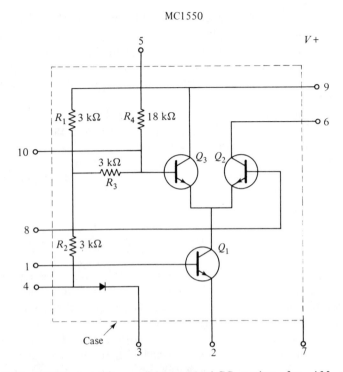

Figure 8.15 RF-IF amplifier, with good AGC action for AM receivers (MC1550); with high-frequency input at terminal ① transistors Q_1 and Q_2 form a cascade circuit, while AGC voltage at terminal ⑤ provides automatic-gain-control action at the base of transistor Q_3. (*Motorola Semiconductor Products, Inc.*)

output stage, until gain has been reduced nearly to zero. Because some leakage always exists, a practical AGC range of about 300:1 (or 50 dB) can be achieved with the monolithic AGC RF-IF amplifier. Because the current-source input stage is looking into two emitters whose parallel impedance remains constant as the AGC signal varies, the monolithic device gives a more constant input impedance than the discrete-component version.

AM-IF Strips with AGC

Following the same large-scale integration trends as in FM-IF strips, devices such as the *National LM 172/372* (Figs. 8.16 and 8.17) include sufficient gain to replace all stages in the conventional strip, and also provide self-contained AGC along with a built-in detector with characteristics suited to the AGC requirements. The *LM 172/372* requires a single-lumped external band-pass filter, because all its internal functions must be direct coupled. Gain variation is

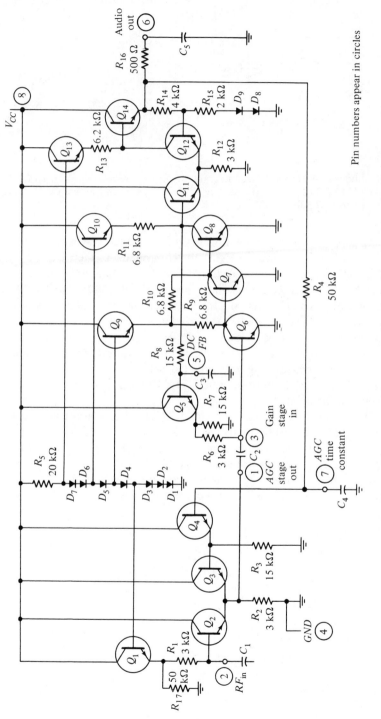

Figure 8.16 AM-IF strip (LM172/372), a receiver subsystem, providing a high gain broadband amplifier (suitable for IF with an external filter) and, in addition, includes excellent AGC action and an active detector. (*National Semiconductor*)

173

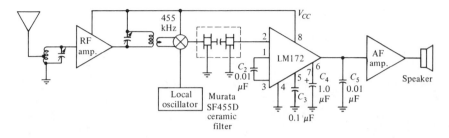

Figure 8.17 Superheterodyne block diagram of an AM receiver circuit, using LM172/372, as an IF strip with AGC action and active detector. (*National Semiconductor*)

achieved by a pair of emitter followers, which act like diodes, one in series with the input and another shunting the signal to RF ground. As their balance is varied, less signal reaches the output of the AGC section and more is shunted to ground, thereby giving an AGC operating range of about 1000:1, or 60 dB. Gain is realized by a set of direct-coupled common-emitter amplifiers, which are forced to remain linear by use of a dc feedback loop (dc feedback loops were discussed in the previous section on FM-IF strips). The direct-coupled detector is far more complex than the simple diode detector used in the discrete AM receiver; it is a feedback amplifier with the detector diode enclosed by a negative-feedback loop, and this accurately sets its operating characteristics without the use of the usual tuning.

An example of an elaborate monolithic AM receiver subsystem is the *Signetics NE546* chip. Referring to the block diagram (Fig. 8.18) of the unit, the chip contains an RF amplifier, IF amplifier, mixer, oscillator, AGC detector, and a voltage regulator. The application of the NE 546 in an AM receiver is illustrated in Fig. 8.19. A dual-ganged variable capacitor is employed for tuning. Winding data for the antenna, RF, oscillator, and IF coils are also included in the figure.

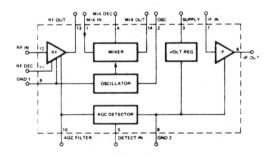

Figure 8.18 Block diagram of the NE546 chip, an AM receiver subsystem. (*Signetics*)

AM RADIO (Capacitor Tuned)

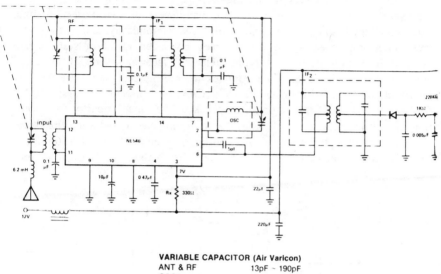

VARIABLE CAPACITOR (Air Varicon)
ANT & RF   13pF ~ 190pF
OSC   12pF ~ 80pF

ANTENNA COIL
10mm φ < 120mm Ferrite Antenna

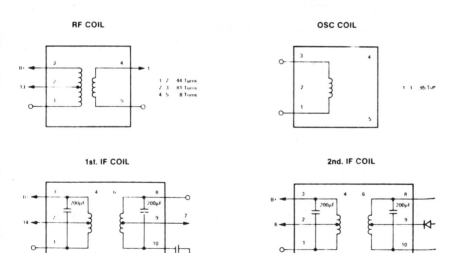

Figure 8.19 Using the NE546 chip in an AM receiver. (*Signetics*)

175

8-6 TELEVISION APPLICATIONS

While the critical high-power and high-voltage sections of television receivers are likely to remain in discrete form for many years, at least until monolithic LIC technology is able to provide such functions economically, many of the low-level functions, especially in the complex color-demodulation systems, are available in monolithic form. A typical television-receiver functional block diagram appears in Fig. 8.20. Four complex monolithic subsystems are used. A television video-IF subsystem, the *RCA CA3068*, contains all these blocks; first, a video-IF/AGC generator for feedback to the tuner, circuitry for derivation of the 4.5-MHz sound IF from the composite signal, and a multiple-stage video IF detector and first video amplifier. A second IC, the *CA3065*, is a sound-IF, sound detector, and audio preamplifier. A third IC, the *CA3066*, pro-

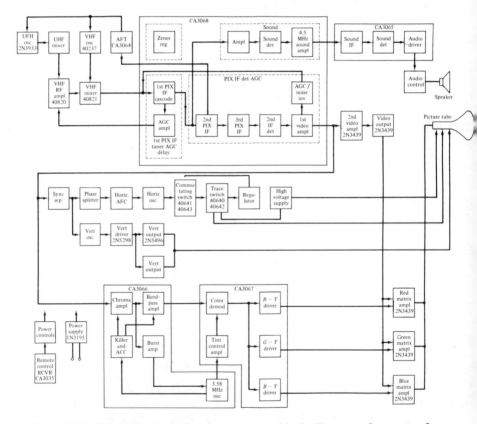

Figure 8.20 Typical color television receiver block diagram; shown are four subsystems, as follows: CA3068, video-IF system; CA3065, sound amplifier section; CA3066, chroma amplifier section; CA3067, chroma demodulator section. (*RCA Solid-State Division, File No. 467*)

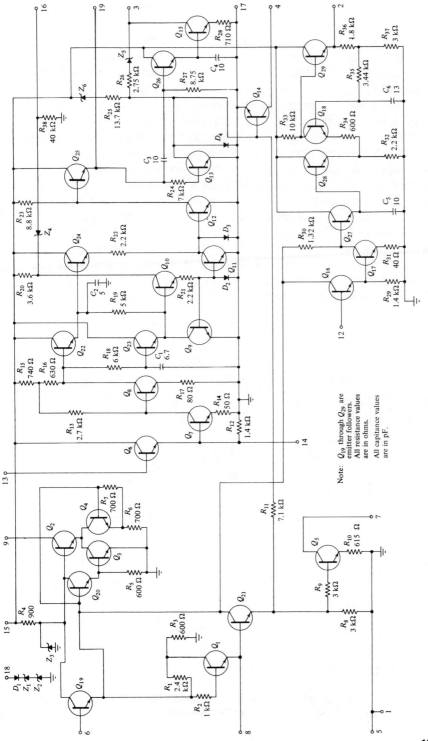

Figure 8.21 Simplified schematic diagram of the CA3068 video-IF system. (*RCA Solid-State Division, File No. 467*)

Note: Q_{19} through Q_{29} are emitter followers.
All resistance values are in ohms.
All capitance values are in pF.

177

vides chroma information through its chroma amplifier, band-pass amplifier, burst amplifier, an automatic chroma control (ACC), and a crystal-controlled 3.58-MHz oscillator, which is locked to the color subcarrier of an incoming signal. The fourth IC, type *CA3067*, performs the critical color-demodulation function, splitting the red, green, and blue phospor-dot information channels for external amplification and distribution to the picture tube.

While a detailed examination of the circuits used in the block diagram of Fig. 8.20 is best left to applications literature available from the manufacturer, some observations can be made from the circuit schematics shown in Figs. 8.21, 8.22, and 8.23. First, it is apparent that a number of external components are still required, being primarily tuned and bypass elements, which are not easily integrated. Second, the complexity of the circuits and the limited number of pins have made these subsystems useful for only a single purpose, rather than being universally applicable types of LICs. To be economical in manufacture, such highly specialized circuits must have very large markets, and the television industry satisfies this criterion. Third, the circuit techniques used in these complex devices are basically the same as have been discussed with respect to single-

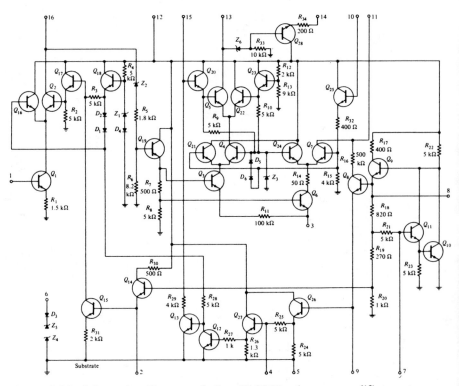

Figure 8.22 Schematic diagram of the CA3066, chroma amplifier system. (*RCA Solid-State Division, File No. 466*)

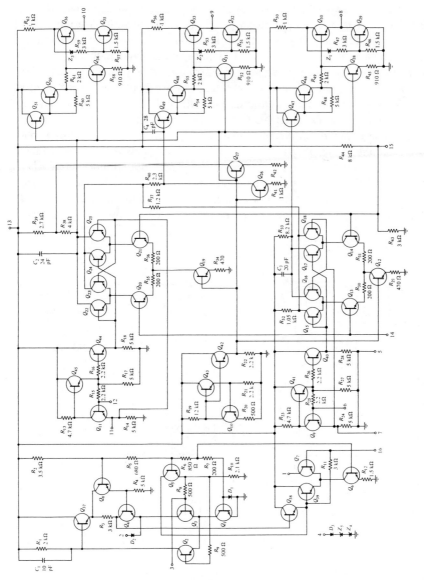

Figure 8.23 Schematic diagram of the CA3067, chroma-demodulator section. (*RCA Solid-State Division, File No. 466*)

179

function FM, AM, and general-purpose LICs and include current sources, differential AGC systems, multipliers, extensive direct coupling, and so on. In other words, complex television circuits are simply logical extensions of existing LIC technology.

The degree to which these LICs can be called large-scale integration (or LSI) is limited primarily by the economics of packaging with large numbers of pins, by the difficulty in providing RF isolation, and by the higher power dissipation that even more consolidated circuits would require. Finally, it is apparent that the reason for using such circuitry cannot be reduction in size, since the picture tube and associated power circuits dominate the chassis of most television receivers. The main advantage of LICs in television receivers is modularity, that is, the use of modules to allow easy replacement of entire defective functions, as opposed to the replacement of only the active devices, thus increasing the likelihood that a repair can be made without extensive troubleshooting.

The *Motorola MC1391P* (for positive flyback inputs) and the *MC1394P* (for negative flyback inputs) are examples of monolithic TV horizontal processors. These chips contain the low-level horizontal sections, including the oscillator, phase detector, and predriver, required for a horizontal loop. A typical

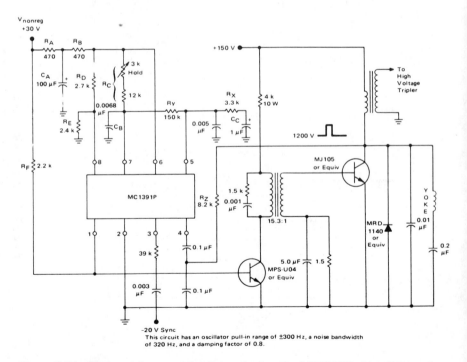

This circuit has an oscillator pull-in range of ±300 Hz, a noise bandwidth of 320 Hz, and a damping factor of 0.8.

Figure 8.24 Horizontal sweep circuit employing the MC1391P chip. (*Motorola Semiconductor Products, Inc.*)

application circuit for the chip is shown in Fig. 8.24. A television vertical proces-sor also is available (*MC1393*). A block diagram of the device in a typical ap-plication circuit is provided in Fig. 8.25.

An example of a monolithic television automatic fine tuning circuit is the *RCA CA3139* chip. The circuit provides an AFT voltage and an amplified 4.5-MHz intercarrier sound signal. When connected to an output of an IF amplifier (Fig. 8.26), the CA3139 provides signal amplification and detection needed to generate the AFT correction signal for the television tuner. It also mixes the video and sound IF carriers and amplifies the resultant 4.5-MHz intercarrier sound signal.

General-Purpose Communication Applications

While the broadcast AM, FM, and TV receiver functions discussed are respon-sible for the bulk of actual communication LIC usage, there remains a vast field of communication applications that is served by a number of general-purpose functions. The applications include *two-way radio*, *navigation*, *telemetry*, *data transmission*, *remote controls*, *alarms*, and other general areas involving the transmission of signals from point to point.

Rather than examine each application separately, the following discussion will concentrate on several *general circuit functions*, which are common to many kinds of communication applications.

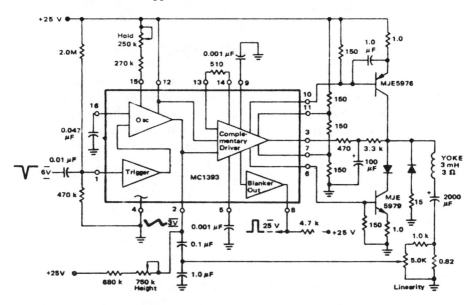

Figure 8.25 Vertical sweep circuit employing the MC1393 chip. (*Motorola Semiconductor Products, Inc.*)

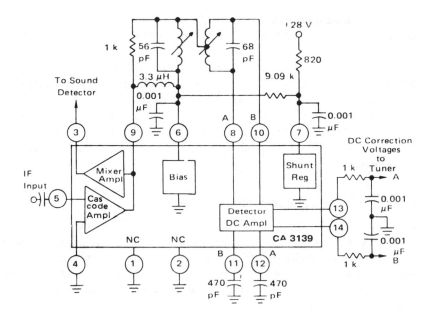

All resistor values are in ohms.

Figure 8.26 Block diagram and typical application of the CA3139. (*RCA Solid-State Division*)

8-7 MIXERS, MODULATORS, AND MULTIPLIERS

A basic requirement in communication systems is to obtain an *output that is proportional to the product of two input signals*. Theoretically, the process of signal multiplication in the time domain is the same as frequency addition and subtraction in the frequency domain—just two ways of looking at the same actual operation. An *RF mixer* can be considered as a multiplier:

$$v_o = k e_{in_1} e_{in_2}.$$

Without dwelling on the mathematical theory, the result of mixing (or multiplying) two frequencies (f_1 and f_2) may be expressed as follows:

$$f_{\text{output}} = Af_1 + Bf_2 + C(f_1 + f_2) + D(f_1 - f_2).$$

That is, the output of such a mixer (or multiplier) contains not only components of both input frequencies (f_1 and f_2), but also the *sum and differences* of the input frequencies.

Similarly, an examination of various forms of amplitude modulation (AM) will show that the modulated output contains the same combinations of two input frequencies. If the inputs to an amplitude modulator are f_1 (RF carrier) and f_2 (audio modulation), the output will contain both input frequencies, as well as their sum and differences, called *sidebands*. In communication equipment, the undesired frequency components are filtered out, leaving only the carrier or carrier-and-sideband information frequencies to be transmitted. In all cases, including AM (carrier plus both sidebands), DSB (both sidebands with carrier filtered out), or SSB (single sideband, where only one sideband remains after filtering), *the necessary modulating process is one of multiplication of two inputs* (also called *mixing*).

Multiplier Types

A structure found in many communication LICs is the *tree-type multiplier*, typified by the *Motorola MC1596* (Fig. 8.27). It is a double-differential amplifier with its outputs "cross connected." Recalling the differential-amplifier AGC circuit (Section 8-5), it is a form of multiplier that produces an output equal to the product of an input signal and a dc voltage. In the AGC amplifier an increase (or decrease, if negatively connected) in dc AGC voltage causes an approximately proportional increase in the signal at the output. In the tree-type multiplier, two such differential circuits are balanced, so that the components of the two input frequencies are canceled at the output, leaving only the sum and difference frequencies. Such a structure is sometimes referred to as a *double-balanced mixer*, and is used in the generation of AM, double-sideband, and single-sideband modulation.

Product Detector: The same structure of double-differential amplifier, when driven by a beat-frequency oscillator (BFO), is used in the receiving-detector section of SSB receivers; the BFO frequency, when added to (or subtracted from) the SSB input, produces a component that is the original audio-modulating frequency. The multiplier nature of this type of SSB detector is shown by its name: *product detector*.

Phase Detector: If all output components except dc are filtered out, the tree multiplier becomes a *phase detector*. If two equal frequencies having a phase difference are used as inputs, a dc output is obtained proportional to the phase shift. If the phase shift is zero, the dc output is zero. This function is important in phase-locked loops, an increasingly important topic to be discussed later.

Quadrature FM Detector: If an *LC* phase-shift network is placed between the input signal and one multiplier input, with the input signal directly connected to the other multiplier input, a phase difference results that varies with input frequency. Thus the tree multiplier becomes a *quadrature FM detector*, which is used, for example, in such television sound IF detectors as the *Sprague ULN2111*. The advantage of such a detector is the requirement only for a simple external *LC* resonant circuit, rather than for a carefully constructed discriminator transformer.

The basic circuit of the *MC1596*, shown in Fig. 8.27, is not strictly a linear

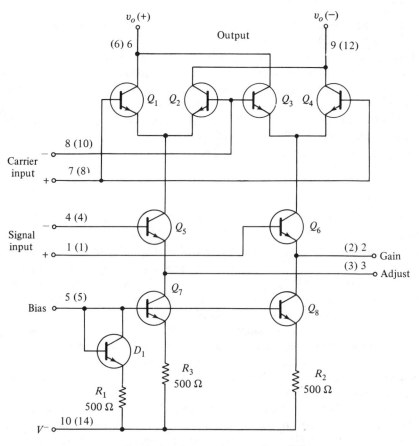

Pin numbers in parentheses apply to *DIP*.

Figure 8.27 Double-balanced mixer-modulator (MC1596): the balanced output is the product of the carrier and signal inputs producing their sum and difference frequencies, with internal carrier suppression, as in single-sideband (SSB) and other modulation applications. (*Motorola Semiconductor Products, Inc.*)

multiplier; a nonlinear distortion of the "upper" input signal results from the nonlinear diode characteristics of the transistor base-emitter junctions; the circuit is nonetheless useful in the functions listed previously, because the spurious frequency components resulting from the nonlinearity are usually filtered out. In *analog computation*, however, a precisely linear product of two inputs is desired. For a linear product, the tree structure has been compensated by using a pair of diodes to cancel the nonlinearity of the multiplier's diode characteristics; this is the type *MC1595, four-quadrant analog multiplier* (Fig. 8.28). The desig-

Circuit schematic

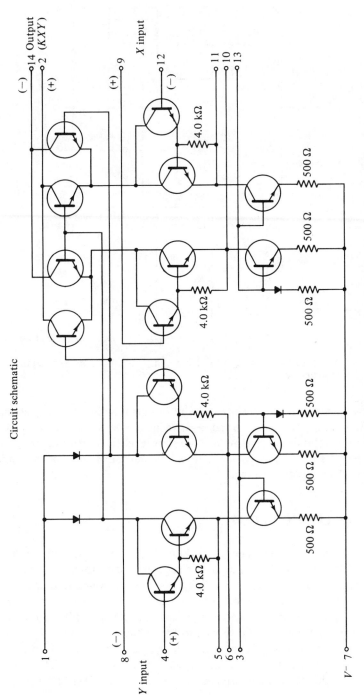

Figure 8.28 Four-quadrant multiplier (MC1595): provides a linear product of X and Y inputs, for analog-computation applications, when used in conjunction with an OP AMP. Typical applications are detailed on MC1595L data sheet. (*Motorola Semiconductor Products, Inc.*)

185

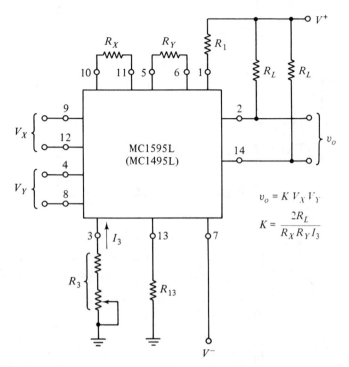

Figure 8.29 A multiplier using the 1595 chip. (*Motorola Semiconductor Products, Inc.*)

nation *four quadrant* refers to the capability of operating in all four quadrants of a graph, in which both positive and negative input 1 are shown on the vertical axis, and positive and negative input 2 are shown on the horizontal axis, thus dividing the possible output polarities into four quadrants. A basic multiplier using the 1595 chip is given in Fig. 8.29.

8-8 PHASE-LOCKED LOOP

A block diagram of a basic phase-locked loop (PLL) is illustrated in Fig. 8.30. The main elements of a PLL are a phase comparator (detector), low-pass filter, and a voltage-controlled oscillator (VCO). An output, which is proportional to the difference between the output frequency f_o and the input frequency $f_i (f_o \pm f_i)$, is provided by the phase comparator. The low-pass filter allows only the difference frequency, $f_i - f_o$, to pass to the input of the VCO. Accompanying the difference frequency is an error voltage, v_e. A typical plot of the error voltage as a function of the phase difference θ between f_i and f_o is illustrated in Fig. 8.31. Note that for a phase difference of 90° the error voltage is zero.

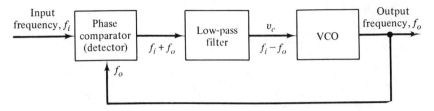

Figure 8.30   Block diagram of a basic phase-locked loop (PLL).

The VCO is an astable (free-running) multivibrator with frequency-determining resistors R connected to the error voltage, as in Fig. 8.32(a). The output of an astable multivibrator is basically a square wave. As v_e increases, timing capacitors C charge more rapidly and the output frequency increases. Conversely, as v_e decreases, capacitors C charge more slowly and the frequency decreases. A plot of output frequency f_o as a function of error voltage v_e is plotted in Fig. 8.32(b). (More will be said about the VCO in Chapter 12.)

Assume that input frequency f_i is different from the free-running frequency, f_o, of the VCO. Further, assume that f_i is within the operating range of the VCO. At frequency f_i then, the frequency of the VCO will change abruptly from f_o and generate a new frequency equal to f_i. The VCO is said to be *locked* to frequency f_i. If the frequency goes outside the operating range of the VCO, the frequency of the VCO will return abruptly to f_o.

Two important terms used to characterize the performance of a VCO are *tracking* and *capture ranges*. The tracking range, also called the *lock-in range*, is the range of input frequency under which a PLL, once locked, remains locked. Capture range, also called the *acquisition range*, is the range of input frequency which permits an initially unlocked PLL to become locked. The capture range is narrower than the tracking range.

FM Demodulator

Suppose that the input frequency is an FM signal. Because a phase-locked loop tries at all times to force the voltage-controlled oscillator to equal the input

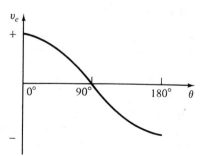

Figure 8.31   A plot of error voltage (v_e) versus the phase difference (θ) between the input (f_i) and output (f_o) frequencies.

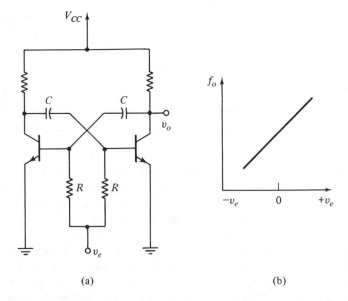

(a)   (b)

Figure 8.32   Basic voltage-controlled oscillator (VCO): (a) circuit; (b) plot of oscillator frequency (f_o) versus the error voltage (v_e).

frequency, the VCO will track the unknown input frequency at a rate limited by the maximum rate of change of the dc error voltage of the VCO. Assume that there is a linear relationship between the error voltage and the operating frequency of the VCO. Obviously, if the reference frequency varies, the error voltage will be a dc voltage that is proportional to the frequency variation. If the input is a 10.7-MHz FM-IF signal that varies ±75 kHz, and it takes a 1-V change in the error voltage to produce a 100-kHz change in the VCO frequency, then the error voltage will be a demodulated replica of the frequency modulation of the input signal, giving ±750 mV of audio output. FM demodulators using phase-locked loops offer the advantage of not requiring external tuned elements, since the VCO can be made with integrated RC elements.

FSK Demodulator

Low-speed data sets, such as a teletypewriter, use a modulation scheme called *frequency-shift keying* (FSK). In this method, two signals of slightly different frequencies and constant amplitude are used to represent 0 and 1 logic levels. Let 1,070 Hz represent a logic 0 and 1,270 Hz a logic 1. Assume that the PLL is locked to one of these frequencies; the output of the PLL will then swing between two dc levels for logic 0 and 1. An example of a monolithic phase-locked-loop FSK demodulator is the *Raytheon XR-2211* chip, shown connected for FSK decoding in Fig. 8.33. For a 300-baud rate and frequencies of 1,070 Hz and

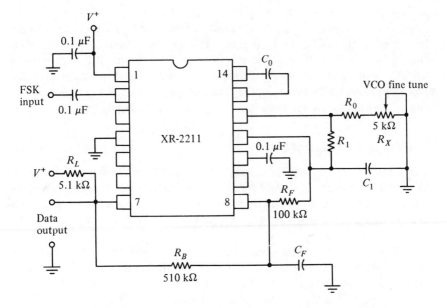

Figure 8.33 The XR-2211 chip used for FSK decoding. (*Raytheon Semiconductor Division*)

1,270 Hz, C_o= 0.039 uF, C_F = 0.005 uF, C_1 = 0.01 uF, R_0 = 18 kΩ, and R_1 = 100 kΩ.

AM Demodulator

A phase-locked loop functioning as a synchronous AM detector is shown in Fig. 8.34. The PLL locks on the carrier of the AM signal so that the VCO output has the same frequency as that of the carrier but no amplitude modulation. The demodulated AM is then obtained by multiplying the VCO signal with the

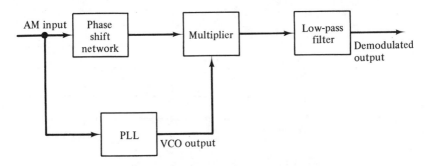

Figure 8.34 A PLL used as a synchronous AM detector.

modulated input signal and filtering the output to remove all but the difference frequency component. When the frequency of the input signal is identical to the frequency of the VCO, the loop goes into lock with these signals 90° out of phase [Fig. 8.31(b)]. If the input is now shifted 90° so that it is in phase with the VCO signal and the two signals are mixed in a second phase comparator, the average dc value (difference frequency component) of the phase comparator output will be directly proportional to the amplitude of the input signal.

Frequency Synthesizer

The frequency synthesizer is a frequency source whose different frequencies have the high stability of a single crystal-controlled reference frequency. A block diagram of a frequency synthesizer is illustrated in Fig. 8.35. Binary counters act like dividers. For example, a 3-stage counter divides by 3, a 4-stage counter divides by 4, and so on.

The frequency of the crystal oscillator, f_r, is divided by M; hence, one input to the phase comparator of the PLL is f_r/M. Another counter, which divides by N, is in series with the VCO. For the loop to lock, both inputs to the phase comparator must have identical frequencies. For this to be true, the VCO frequency (which is also the output frequency, f_o, of the synthesizer) is

$$f_o = \frac{f_r N}{M}.$$

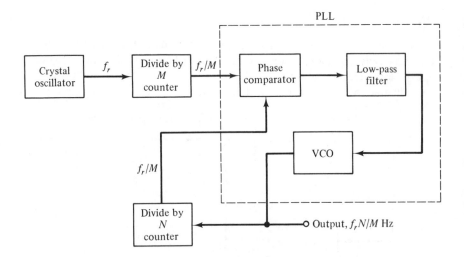

Figure 8.35 Block diagram of a basic frequency synthesizer.

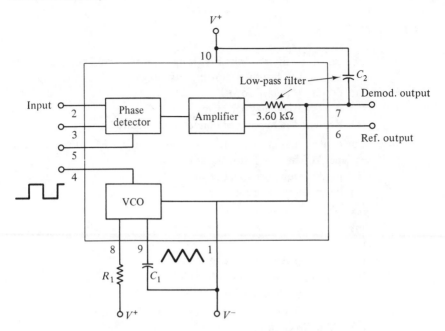

Figure 8.36 Block diagram of the SE565 phase-locked loop. It covers the frequency range of 0.001 Hz to 500 kHz. (*Signetics*)

An example of a monolithic PLL, the *Signetics SE565*, is shown in block diagram form in Fig. 8.36. It covers the frequency range from 0.001 Hz to 500 kHz. The center frequency, f_o, of the VCO is determined by R_1 and C_1 and is expressed by

$$f_o = \frac{1.2}{4\,R_1 C_1} \text{ Hz.}$$

The lock-in range, f_L, is

$$f_L = \pm \frac{8\,f_o}{V_{CC}} \text{ Hz}$$

where V_{CC} is the dc supply voltage. The capture range, f_C, is

$$f_C = \pm \frac{1}{2\pi} \sqrt{\frac{2\pi f_L}{36{,}000\,C_2}} \text{ Hz}$$

QUESTIONS

8-1. Compared to discrete circuits using multiple transistors, the communication LIC has the following advantage:
(a) Better optimized for RF
(b) Is more frequency selective
(c) Reduces the number of connections required

8-2. For control action of RF-IF amplifiers used for FM, the LIC (such as type 703) has the following advantage:
(a) Better frequency selection
(b) Better limiting action
(c) Higher frequency response

8-3. Integrated-circuit devices for communication circuits are mainly used
(a) For integrated subsystems
(b) For improvement in cross-modulation
(c) For improvement in frequency response

8-4. The use of LICs in communication circuits is mainly in
(a) FM radio circuits
(b) TV circuits
(c) FM and TV circuits

8-5. Describe the behavior of a VCO in PLL circuits.

PROBLEMS

8-1. In Fig. 8.35, $f_r = 1$ MHz. For VCO frequencies of 0.1, 2, 5, and 10 MHz, determine the ratio N/M.

8-2. For the SE565 phase-locked loop, what should the $R_1 C_1$ time constant be for $f_o = 0.001$ Hz?

8-3. Repeat Problem 8-2 for $f_o = 500$ kHz.

8-4. If $f_o = 0.001$ Hz, determine the lock-in range f_L for the SE565; assume $V_{CC} = 12$ V.

8-5. Repeat Problem 8-4 for $f_o = 500$ kHz.

8-6. Based on the results obtained in Problems 8-4 and 8-5, calculate the capture range, f_C.

Chapter 9

Regulators and Control Circuits

9-1 POWER CONTROL WITH MONOLITHIC CIRCUITS

While the internal power dissipation capability of a monolithic LIC is limited by the difficulty of removing substantial amounts of heat from a very small package, the precision and complexity of the circuit make it an effective control element for external power transistors and also for thyristors (SCRs, or silicon-controlled rectifiers) and other devices that operate at high power levels, but are themselves imprecise. Power-control LICs are achieving widespread use in home appliances, lamp dimmers, motor controls, thermostats, safety circuit breakers, automotive applications, and much more. In all these control applications the LIC *performs timing, regulation, or overload protection* or, at times, a combination of these functions. In low- and medium-power uses, the LIC may contain the power-handling elements internally, but in many cases inexpensive external elements are used.

The use of external power components may be required for one of three reasons: power dissipation, maximum current handling, or maximum breakdown voltage limitations of the LIC fabrication process. The most economical, easily reproduced LICs are those operating at currents of less than 1 A, voltages of less than 40 V, and package dissipations of less than 1 W. Each of these parameters can be increased by a factor of 10 with current technology, but trade-offs must always be considered, especially when very inexpensive external power elements could be used instead.

9-2 VOLTAGE REGULATORS

Many linear, and nearly all digital, systems require a constant, precisely regulated power supply voltage, or several such voltages. Certain LICs provide such regulation internally for use in the critical sections of their own circuitry. Certain systems use *local regulation*, with a separate voltage regulator for each section of the system, not only to provide constant voltages, but also under operating conditions to prevent variations in power consumption by one section from being transmitted back along the common supply lines to other sections.

Voltage regulators may be classified, on the one hand, by the method of achieving regulation, as in *linear regulators* or *switching regulators*, and, on the other hand, by their means of connection, as in *series regulators* or *shunt regulators*. A linear voltage regulator contains five basic elements: (1) a precision voltage reference, (2) a sampling element, (3) a comparator, (4) an error amplifier, and (5) a power control element. The *three main elements* are the voltage reference and the error and control amplifiers.

The precision voltage reference provides a stable voltage level. A zener diode is generally used for this purpose. Sampling or monitoring of the output voltage is achieved with a sampling element, typically a voltage divider. The comparator compares the sample of the output voltage with the voltage reference. The difference, called the *error voltage*, is amplified by the error amplifier. A control element, which may be in series with the load in the series regulator or in parallel with the load in the shunt regulator, is driven by the error amplifier.

A block diagram of a series regulator is illustrated in Fig. 9.1. The basic elements are connected in a negative-feedback loop. Assume that, for a constant load current I_L, input voltage V_{in} increases. Load current I_L and output voltage V_o, therefore, will tend to increase, resulting in an increase of the sampling voltage to the comparator. The difference in the sampling and reference voltages rises and the error voltage increases. This causes the control element to increase its internal resistance, thereby increasing the voltage drop across it. This increase in voltage offsets the increase in the input voltage, and the output remains essentially constant. If the input voltage decreases, the reverse occurs. In this case, the error voltage is reduced and the voltage drop across the control element decreases. Voltage regulation also is obtained for changes in load resistance or load current.

Two important criteria for establishing the performance of a voltage regulator are the *input* or *line regulation* and *load regulation*. *Input regulation* is the percentage change in the output voltage, $\Delta V_o/V_o$, for a change in the input voltage, ΔV_i:

$$\text{Input regulation} = \frac{\Delta V_o/V_o}{\Delta V_i} \times 100\%.$$

Voltage V_o is the nominal value of the output voltage.

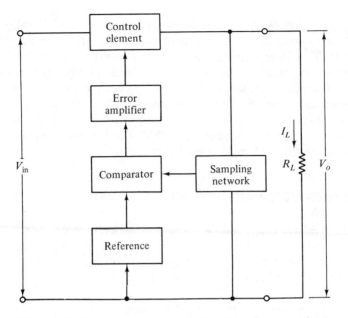

Figure 9.1 Block diagram of a basic series voltage regulator.

Load regulation is the change in the output voltage for a change in load current:

$$\text{Load regulation} = \frac{V_{o2} - V_{o1}}{V_{o1}} \times 100\%,$$

where V_{o2} is the voltage at I_{L2} and V_{o1} is the voltage at I_{L1}. Current I_{L2} is less than current I_{L1}.

The majority of LIC voltage regulators are specifically designed for one supply polarity, either positive or negative, although universal, dual-polarity regulators also are available. Many LIC regulators also include *current limiting*, which may be of either the *constant-current* or *foldback-limiting* types, and *overload protection*.

Linear versus Switching Regulators

There are two basic types of regulators: *linear* and *switching regulators*. In the linear regulator, described in the preceding paragraphs, there is an active element whose resistance may be continuously varied by an amplified error voltage to suit the current demand. In the switching regulator a saturating, ideally lossless switch, is turned on and off at a rate that delivers the desired average current in periodic pulses to the load.

The switching regulator is a much more efficient circuit than the linear type, since the switching element theoretically dissipates no internal power in either the on state, with no voltage across it, or in the off state, with no current through it. The linear regulator must dissipate power in the control element continuously. Practically, however, a greater number of linear regulators are in use. First, there is a finite time in every switching cycle during which the switching element makes the transition between on and off; during this time, dissipation is substantial. Second, many loads cannot tolerate the periodic delivery of energy, even if smoothed by a capacitor. Third, the switching regulator could be a troublesome source of transients and RF radiation.

9-3 SERIES VERSUS SHUNT REGULATORS

In a series regulator (Fig. 9.1), the variable-resistance, or switching control, element is placed between the raw, unregulated supply and the load. As the load demands more current, the series resistance is decreased, so that the voltage divider formed by the series control element and the load gives the desired output voltage.

In a shunt regulator the unregulated supply is either a high-impedance current source or an unregulated voltage source with a substantial fixed series output impedance. Referring to Fig. 9.2, the control element Q_1 is shunted across the load and is increased in impedance as load current demand increases. The net current, $I_i \simeq I_1 + I_L$, into the parallel combination of load and control elements remains constant; consequently, the voltage across the load also remains constant.

A series regulator is desirable in applications where load-demand current varies from zero to some fixed maximum; dissipation in the control element is zero when load current is zero. A shunt regulator is desirable when load current

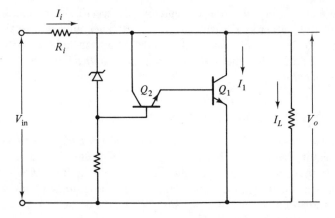

Figure 9.2 A basic shunt-type voltage regulator.

varies from a finite minimum value to a finite maximum value; dissipation in the control element is low or zero when load current is at its maximum, but unnecessary power is then consumed in the fixed source resistance, R_i, of the unregulated supply. Most LIC regulators are designed for use as series regulators.

9-4 CURRENT LIMITING

Although not a necessary part of a voltage regulator, the complex construction possible in monolithic technology permits most LIC regulators to provide a form of current-limiting action. Usually, a small resistance is included in series with the load. The load current produces a small but predictable voltage across this *sensing resistor*, which is amplified and used to trigger a clamping circuit to defeat the negative-feedback loop that ordinarily controls the circuit. If load current continued to increase without this limiting action, the negative-feedback loop would attempt to provide the required large currents, ultimately destroying the control element, unless other resistances in the circuit or in the unregulated supply were large enough to prevent damage. *The current limiter, however, allows direct short circuits, in most cases, without permanent destruction.*

The simplest limiter is one in which a threshold is reached above which regulator output current, rather than voltage, is held constant. Thus a direct short circuit would have the same limited current through it as would a load slightly heavier than that permitted by the limiter. Heavy stress is placed upon *series regulators* in this kind of limiting, since they may have maximum voltage and constant, maximum current applied simultaneously, permitting a large but controlled power dissipation.

In the *foldback limiter*, when maximum load current is exceeded, the output voltage is folded back or reduced to a lower value, or even zero. Although this provides much safer operation, it does permit power to be shut off in case of fault from a load that may otherwise require continuous power. Such a fault may be a very short, momentary one; nevertheless, the regulator will stay shut off until externally reset.

9-5 REPRESENTATIVE EXAMPLE OF A VOLTAGE REGULATOR   *723*

The three main elements of a basic regulator (such as the widely used μA723) are shown in functional-equivalent form in Fig. 9.3(a). Here the reference-voltage function (V_{ref}) is initially determined by the 6.2-V zener diode; this is fed to one of the inputs of the error amplifier, while the other input receives a portion of the output, resulting in an amplified error that controls the output of the series-pass (control) transistor.

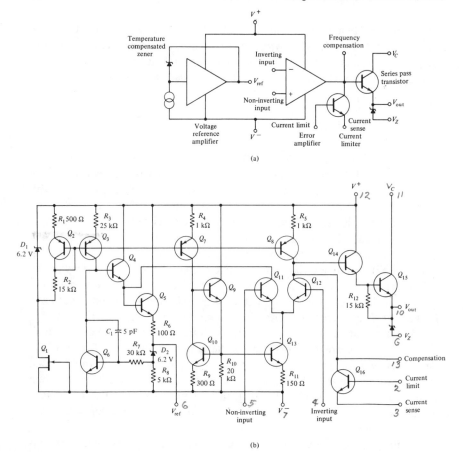

Figure 9.3 Voltage regulator (μA723): (a) functional equivalent, with main elements—reference-voltage, error-amplifier, and control (series-pass) element; (b) schematic-equivalent circuit. (*Fairchild Camera and Instrument Corporation*)

Reference Voltage

This element is the standard against which the regulated output is compared. Accordingly, the accuracy of the regulated output is highly dependent upon the accuracy of the reference. Since monolithic processes are subject to wide variation, the absolute value of the reference is often hard to control, being established by forward or reverse diode breakdown voltages. Usually, a voltage divider external to the LIC is used to set the ratio between the output voltage and the reference, so the actual initial value of the reference is less important than its stability with time and temperature. Since the temperature characteristics of various junction breakdowns are more predictable than their absolute

$LIC = Linear\ Integrated\ Circuit$

values, regulator references are often a series combination of the forward voltage drop and zener breakdown calculated to give minimum temperature drift. In extremely precise regulators, a reference external to the LIC may be used.

Error Amplifier

This element compares the regulated output voltage against the reference voltage, and drives the control element, more or less, to compensate for load-induced variations in output voltage. It is desirable to have a large gain in the error amplifier so that very small differences between desired and actual output voltages will cause large and immediate correction outputs. Very large gains, however, are often the source of high-frequency instability that is due to phase shifts in the feedback loop, which must be corrected by compensation capacitors (as is the case with operational amplifiers). The more rapidly the error amplifier can react, the better; otherwise, high-frequency variations in load impedance cannot be "followed" by a slow-acting error amplifier, and will produce unregulated outputs. It is also important that the error amplifier recover quickly from overload, and also from the initial transient resulting when the regulator is first turned on.

Control Element

This must be capable of handling both the maximum voltage and maximum current to be encountered, and of dissipating the power resulting from a worst-case combination of voltage and current. A fast-acting error amplifier may be worthless if the control element is not also capable of fast response. In the case of switching regulators, fast recovery from switching transients is also a must, as power is dissipated in such regulators during the switching transition period, as well as the on period.

As an example of a general-purpose regulator, the *Fairchild μA 723* has become well known, featuring a highly flexible circuit that can be used in many different ways. The schematic-equivalent circuit is shown in Fig. 9.3(b).

In general, the 723 type is capable of regulating either *positive or negative voltages, fixed or adjustable*, and in either *series or shunt* configuration. Its line regulation is typically 0.01 percent, and it will regulate typically with 0.03 percent against load changes (1 to 50 mA).

When output currents greater than 50 mA are required, it is designed for use with an external *NPN* or *PNP* power transistor. Terminals are shown on the schematic diagram for current limiting (CL) and current sensing (CS), and also provision is made for a sensing-current resistor (R_{SC}) and for a compensating capacitor (COMP). Connections to these terminals are made in various ways for the specific applications described in the next section.

9-6 APPLICATIONS OF THE 723-TYPE REGULATOR

A number of typical voltage-regulator applications are shown in the accompanying figures.

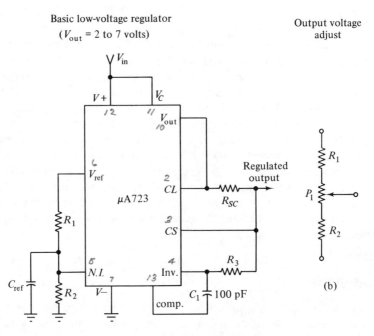

Basic low-voltage regulator
(V_{out} = 2 to 7 volts)

Output voltage
adjust

(b)

Typical performance

Regulated output voltage	5 V
Line regulation (ΔV_{in} = 3 V)	0.5 mV
Load regulation (ΔI_L = 50 mA)	1.5 mV

Note: $R_3 = \dfrac{R_1 R_2}{R_1 + R_2}$ for minimum temperature drift.

For outputs from +2 to +7 volts

$$V_{out} = V_{ref} \left[\frac{R_2}{R_1 + R_2} \right]$$

(a)

Figure 9.4   Basic low-voltage regulator (to +7 volts): (a) for fixed output voltage; (b) for variable output voltage. (*Fairchild μA723 Data Sheet*)

Basic Positive-Voltage Regulation

The first two illustrations, Figs. 9.4 and 9.5, show the connections for positive-voltage regulation. Figure 9.4 is for basic *low-voltage applications* (up to +7 V), which covers the familiar 5-V fixed output widely used in digital-logic circuits. The use of a *potentiometer (P_1) for adjustable voltages* in this range is shown in part (b).

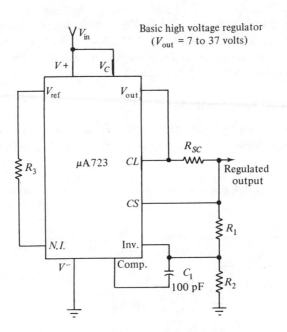

Basic high voltage regulator
(V_{out} = 7 to 37 volts)

Typical performance

Regulated output voltage	15 V
Line regulation (ΔV_{in} = 3 V)	1.5 mV
Load regulation (ΔI_L = 50 mA)	4.5 mV

Note: $R_3 = \dfrac{R_1 R_2}{R_1 + R_2}$ for minimum temperature drift.

R_3 may be eliminated for minimum component count.

For outputs from +7 to +37 volts

$$V_{out} = V_{ref} \left[\frac{R_1 + R_2}{R_2} \right]$$

Figure 9.5 Basic high-voltage regulator (to +37 volts): see text for note on specific voltage-divider values for fixed or variable output voltages. (*Fairchild μA723 Data Sheet*)

Note: Specific values for R_1, P_1, and R_2 for this application (and the applications that follow) are given in the *μA723* data sheet.

The connections for a basic *high-voltage regulator* are shown in Fig. 9.5. Here it will be noted that the junction of the voltage divider (R_1, R_2), which sets the desired output voltage, is now connected to the inverting (INV) terminal of the error amplifier, and thus determines output voltages higher than the reference, as shown in the formula on the diagram. Specific values for the voltage-divider resistors (R_1, R_2) are given in the data sheet up to +28 V fixed, and also for the variable pot (P_1) of Fig. 9.4(b), when adjustable output is desired. This application covers the +15-V *output value* that is popularly used for the positive supply of OP AMPs. (Data for circuit connections and values up to +250 V are also given on the data sheet.)

Using External Pass Transistor

The use of an externally connected power transistor for output currents greater than 150 mA is shown in Fig. 9.6. Here an external *NPN* pass transistor is con-

Positive voltage regulator
(external *NPN* pass transistor)

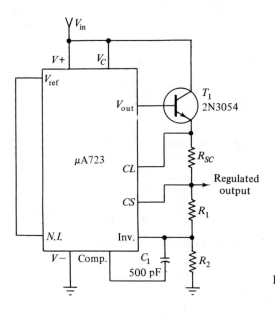

Typical performance

Regulated output voltage   + 15 V

Line regulation (ΔV_{in} = 3 V)   1.5 mV

Load regulation (ΔI_L = 1 A)   15 mV

Figure 9.6   Regulator with external pass transistor: used for output currents exceeding 150 mA (*Fairchild μA723. Data Sheet*)

nected for positive-voltage regulation, and the performance data are given for a +15-V output, regulated for *load changes up to* 1 A. (The data sheet also gives connections and values for using a *PNP* pass transistor in a negative-voltage regulator.)

Negative-Voltage Regulator

The connections for a negative-voltage regulator, shown in Fig. 9.7, involve the use of an external *PNP* transistor with the unregulated input (V_{in}) applied to the collector of this transistor. Using the proper values for the voltage divider

Negative-voltage regulator

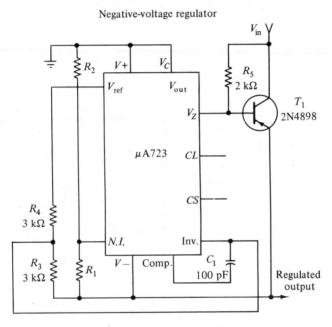

Typical performance

Regulated output voltage	-15 V
Line regulation (ΔV_{in} = 3 V)	1 mV
Load regulation (ΔI_L = 100 mA)	2 mV

For outputs from -6 to -250 volts:

$$V_{out} = \frac{V_{ref}}{2} \left[\frac{R_1 + R_2}{R_1} \right]$$

$$R_3 = R_4$$

Figure 9.7 Negative-voltage regulator: specific values for voltage-divider resistors (R_1, R_2) for fixed output are given on data sheet; see Fig. 9.4(b) for adjustable output connections. (*Fairchild μA723 Data Sheet*)

(R_1, R_2) as given on the data sheet, the performance data for the widely used *negative 15-V supply* are shown on the figure, for both line and load regulation.

Current-Limiting Action

Connections for *current-limiting action of the foldback type* are given in Fig. 9.8. For the typical regulating performance indicated in the figure, the circuit will limit the short-circuit current to 20 mA.

Foldback current limiting

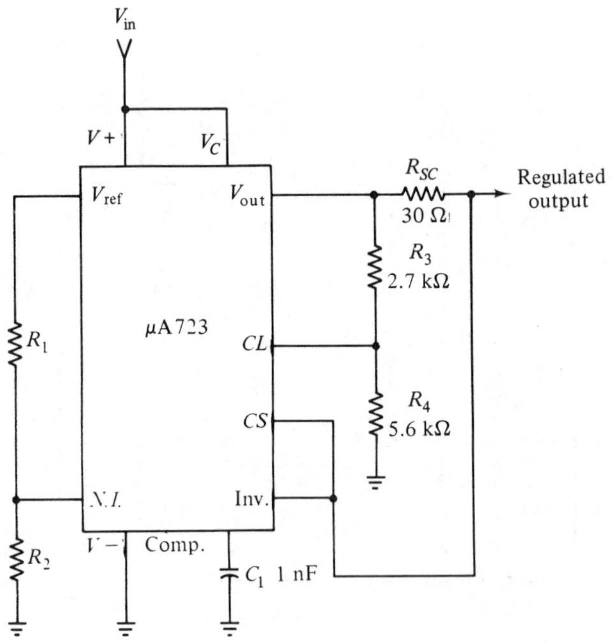

Typical performance

Regulated output voltage	+ 5 V
Line regulation (ΔV_{in} = 3 V)	0.5 mV
Load regulation (ΔI_L = 10 mA)	1 mV
Short circuit current	20 mA

Figure 9.8 Foldback current limiting: with the values shown for the sensing-current resistor (R_{SC}) and at the current-limiting terminal (CL), the short-circuit current is limited to 20 mA. (*Fairchild μA723 Data Sheet*)

Switching-Regulator Action

The connections for switching-regulator action of the 723 type regulator are shown in Fig. 9.9 (for positive output) and in Fig. 9.10 (for negative output). In this type of action, the regulator requires the use of two external transistors, and will handle relatively large changes in input voltage (V_{in}) and up to 2-A change in load current (I_L).

Positive switching regulator

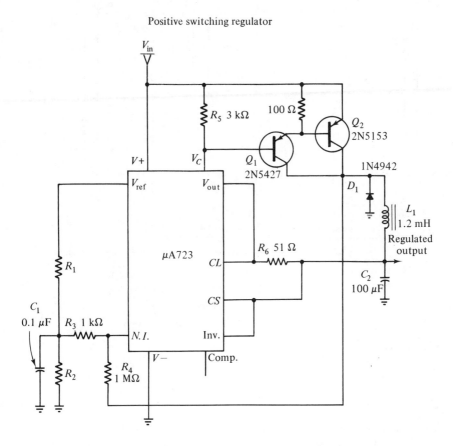

Typical performance

Regulated output voltage	+ 5 V
Line regulation (ΔV_{in} = 30 V)	10 mV
Load regulation (ΔI_L = 2 A)	80 mV

Figure 9.9 Switching-regulator action for positive output voltage. (*Fairchild μA723 Data Sheet*)

Negative switching regulator

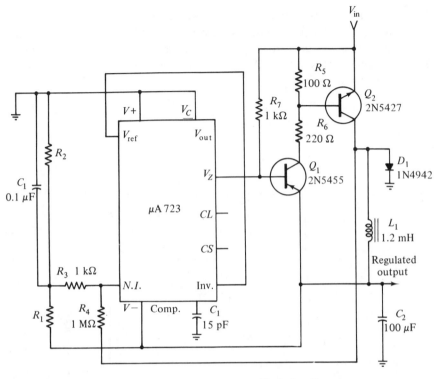

Typical performance

Regulated output voltage	− 15 V
Line regulation (ΔV_{in} = 20 V)	8 mV
Load regulation (ΔI_L = 2 A)	6 mV

Figure 9.10 Switching-regulator action for negative output voltage. (*Fairchild μA723 Data Sheet*)

Shunt Regulator

The connections for applying the type 723 as a shunt regulator are shown in Fig. 9.11. Here the external transistor is connected to the V_Z terminal of the 723, and the regulated output is taken from the collector of this transistor, rather than from the ordinary V_{out} terminal of the device, when used as a series regulator.

Additional applications, such as connections for remote control of the regulator and others, are given in the data sheet for the μA723 type, along with the specific values of the voltage divider to be used for obtaining either fixed or variable output voltages.

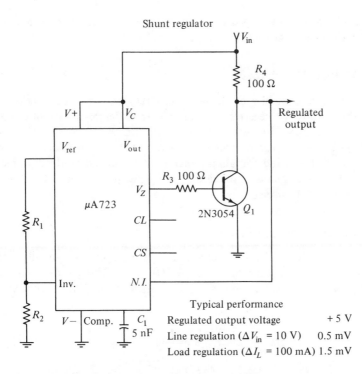

Shunt regulator

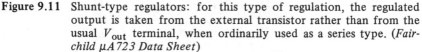

Typical performance

Regulated output voltage	+ 5 V
Line regulation (ΔV_{in} = 10 V)	0.5 mV
Load regulation (ΔI_L = 100 mA)	1.5 mV

Figure 9.11 Shunt-type regulators: for this type of regulation, the regulated output is taken from the external transistor rather than from the usual V_{out} terminal, when ordinarily used as a series type. (*Fairchild μA723 Data Sheet*)

9-7 OTHER REGULATOR TYPES

As representative of voltage regulators that also provide current regulation, mention may be made of the *Motorola* wide-range "floating type," *MC1566/1466;* the *National LM109*, a local 5-V regulator used primarily for digital logic cards (but also capable of current regulation beyond 1 A in an alternative TO-3 power package); and the *LM376*, a positive-voltage regulator in a compact eight-pin mini-DIP package.

A versatile power-control type of LIC is the *CA3094A* programmable power-switch/amplifier. Not only can it function as a monolithic high-current-output device (as stated in Section 7-2), but it can also serve for programmable strobing and gating, making it suitable for control of temperature and motor speed. In the switching mode it can deliver 3 W average (or 10 W peak) to an external load. Its power dissipation (P_D up to T_A = 55°C) is 630 mW (without heat sink), and up to 1.6 W with heat sink.

A variety of power packages are offered for different power levels, as shown in the data manuals of the various manufacturers.

Fixed-Voltage Regulators

There exists a large family of three-terminal fixed voltage regulators that provides either positive or negative regulated voltages. Representative of this group is the *Lambda LAS 1500* series for positive voltages (+5 to +28 V) and the *LAS 1800* series for negative voltages (−2 to −28 V). Both devices regulate at currents up to 1.5 A and contain internal short-circuit and overload protection. Functional block diagrams of the regulators are provided in Fig. 9.12. Each type is housed in a TO-3 package.

Dual-Polarity Tracking Regulators

OP AMPs generally require two voltages for their operation, typically ±15 V. Representative of this group of regulators is the *MC1568/MC1468* dual-polarity tracking regulators that provide balanced positive and negative output voltages

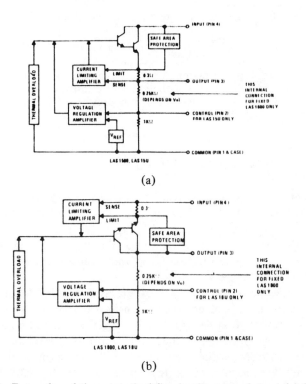

(a)

(b)

Figure 9.12   Examples of three-terminal fixed voltage regulators: (a) for positive voltages; (b) for negative voltages. (*Lambda Electronics*)

at currents up to 100 mA. The current range may be extended with the use of external pass transistors.

Internally, the unit is set for ±15-V outputs. An external adjustment, however, can be used to change both outputs simultaneously from 8 to 20 V. Input voltages up to ±30 V can be applied, and there is provision for adjustable current limiting. The unit is available in three different types of cases (603C, 614, and 632) to accommodate various power requirements.

The application of the MC1568/MC1468 device as a basic 50-mA regulator is illustrated in Fig. 9.13. Using external power transistors, the device used in a ±1.5-A regulator is shown in Fig. 9.14. An example of its use at output voltages from 8 to 14.5 V is provided in Fig. 9.15.

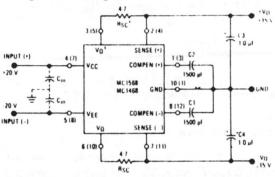

Figure 9.13 The MC1568/MC1468 used as a basic 50-mA regulator. (*Motorola Semiconductor Products, Inc.*)

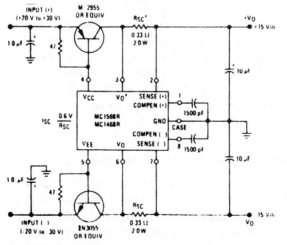

Figure 9.14 Using external power transistors for a ±1.5-A regulator. (*Motorola Semiconductor Products, Inc.*)

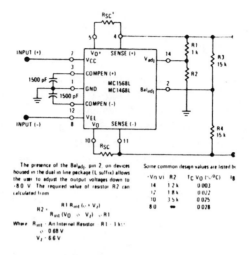

Figure 9.15   The use of the MC1568/MC1468 in an 8- to 14.5-V regulator. (*Motorola Semiconductor Products, Inc.*)

Precision Voltage References

Typical of these devices is the *Beckman Series* 840 temperature-compensated, thin-film hybrid voltage references. The unit accepts an unregulated dc voltage and delivers an accurate positive 10-V dc output voltage over the specified operating temperature range. The output voltage accuracy is ±0.1 percent maximum and the output voltage TC is ±5 ppm/°C. Output short-circuit protection is included and the unit is housed in an eight-pin TO-5 package. The output voltage and temperature stability are realized by the use of thin-film resistors and automatic laser trimming during assembly.

The schematic diagram of the device is given in Fig. 9.16. In operation, pins 1 and 8 must be connected externally as shown on the diagram. Output voltage, V_{out}, is given by $V_{out} = (1 + R_1/R_2) VR$, where $VR = 10$ V. Resistor R_4 sets the current through the zener transistor branch, and R_3 minimizes the OP-AMP output voltage offset errors. Although VR is preset at 10 V ±0.1 percent, an adjustment pin (pin 5) is available to the user for setting the device to a better specification or to a lower voltage (7 to 10 V).

Three-Terminal Adjustable Voltage Regulators

Representative of this group of regulators is the *National Semiconductor LM 117/217/317* series of adjustable three-terminal positive voltage regulators capable of supplying in excess of 1.5 A over an output voltage range of 1.2 to 37 V. They require only two external resistors to set the output voltage. Further, they employ internal current limiting, thermal shutdown, and safe area compensation, making them essentially blow-out proof.

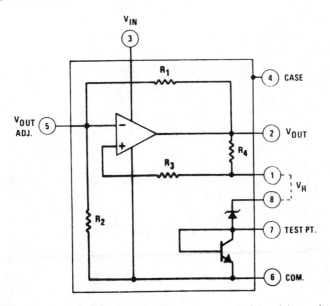

Figure 9.16 An example of a temperature-compensated precision voltage reference (Series 840). (*Beckman Instruments, Inc.*)

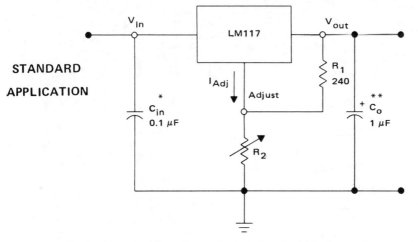

* = C_{in} is required if regulator is located an appreciable distance from power supply filter.

** = C_o is not needed for stability, however it does improve transient response.

$$V_{out} = 1.25 \text{ V } (1 + \frac{R_2}{R_1}) + I_{Adj} R_2$$

Since I_{Adj} is controlled to less than 100 μA, the error associated with this term is negligible in most applications

Figure 9.17 An example of a three-terminal adjustable voltage regulator (LM117). (*National Semiconductor*)

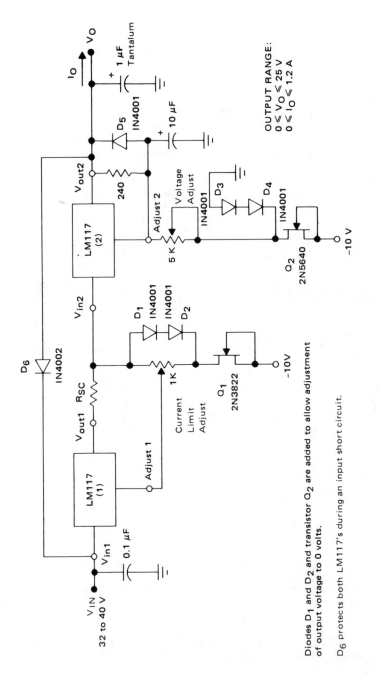

Figure 9.18 The LM117 employed in a laboratory power supply. (*National Semiconductor*)

Diodes D_1 and D_2 and transistor Q_2 are added to allow adjustment of output voltage to 0 volts.

D_6 protects both LM117's during an input short circuit.

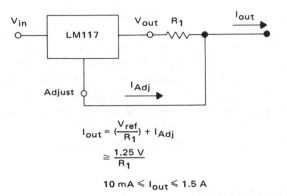

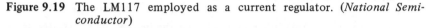

$$I_{out} = (\frac{V_{ref}}{R_1}) + I_{Adj}$$

$$\cong \frac{1.25\ V}{R_1}$$

$$10\ mA \leqslant I_{out} \leqslant 1.5\ A$$

Figure 9.19 The LM117 employed as a current regulator. (*National Semiconductor*)

The LM 117 series serves a wide variety of applications, including local, on-card regulation. This device also makes a simple adjustable switching regulator or a programmable output regulator, or by connecting a fixed resistor between the adjustment and output terminals, the LM 117 series can be used as a precision current regulator. Its use in a standard application is shown in Fig. 9.17. Other applications as a laboratory power supply and a current regulator are illustrated in Figs. 9.18 and 9.19, respectively.

Switching Regulators

An example of a monolithic switching regulator is the *Fairchild µA78S40* device. Referring to the block diagram of Fig. 9.20, the 78S40 consists of a

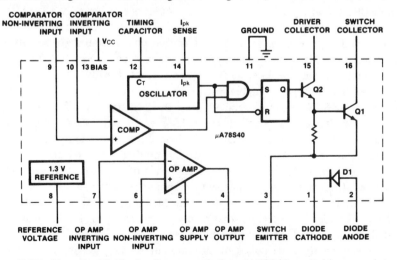

Figure 9.20 Functional block diagram of the µA78S40 switching regulator. (*Fairchild Camera and Instrument Corporation*)

temperature-compensated voltage reference, a duty-cycle controllable oscillator with an active current limit circuit, an error amplifier, high-current, high-voltage output switch, a power diode, and an uncommitted operational amplifier. The device can drive external *NPN* or *PNP* transistors when current in excess of 1.5 A or voltages in excess of 40 V are required. The device can be used for step-down, step-up, or inverting switching regulators as well as for series pass regulators. It features wide supply voltage range, low standby power dissipation, high efficiency, and low drift. It is useful for any stand-alone, low part-count switching system and works especially well in battery-operated systems.

9-8 ZERO-VOLTAGE SWITCHES

In contrast to the control of dc power, which is the field of the voltage-regulator type of LIC, most industrial and household power is from the ac line, usually at 115 V ac, 60 Hz. The control requirements of such line-operated devices generally exceed both the voltage- and current-handling capabilities of the previously described LICs.

In controlling the high-power line-operated loads, such as heaters, valves, and motors, the older methods generally rely on such classic *electrical means* as relays, power rheostats, and variable transformers. The newer *electronic methods* make use of the *silicon-controlled rectifier* (SCR) or thyristor, and other triggered devices, such as the *triac*.

Controlling the SCR

The SCR, a four-layer semiconductor power device, is basically a latch that is either an open circuit or a low-resistance saturated two-terminal device. A third terminal, the gate, is used as a trigger point, requiring relatively little drive current. Many SCR-operated appliances, such as variable speed drills, light dimmers, and the like, derive trigger current directly from the available ac power line through a variable resistor that sets the point on the ac waveform at which the SCR begins to conduct, with the latch turning off when ac supply polarity reverses on the next half-cycle. Simple SCR controls, therefore, permit controlled power to be delivered to the load during a portion of the ac cycle, varying from zero to the full cycle, depending on the control setting.

There are instances in which the triggering of the SCR must be more precisely controlled than is possible with a simple potentiometer, or in which a system control signal is used to drive the SCR rather than a manual control setting. Consider, for example, that very large current load is to be switched, but that only a small SCR is available or economical. The small SCR can probably sustain fairly large pulses of current, but is likely to be damaged if the actual switching is accomplished while the ac-line waveform is at its maximum peak voltage, since this implies maximum power dissipation at the instant of turn on. If the SCR is turned on precisely when the ac waveform is passing through its zero-crossing voltage, no power is dissipated until a later point in the cycle, when the switch is already closed. The *zero-crossing-switch* LIC is a sensitive detector that detects this critical timing, and drives the SCR trigger directly.

Representative Zero-Voltage Switch

A typical zero-voltage switch, the *CA3059*, is shown functionally in Fig. 9.21. The 14-pin DIP package contains the following basic functions:

1. *Limiter power supply,* which allows the LIC to derive its operating power from the high-voltage ac line.
2. *Differential on–off sensing amplifier,* which receives input on–off signals from an external NTC (negative-temperature coefficient) sensor or command signal, and gates ON in the rest of the circuit, in preparation for the next-arriving zero crossing.
3. *Zero-crossing detector,* which monitors the ac line voltage and gives an output pulse when that voltage is zero.

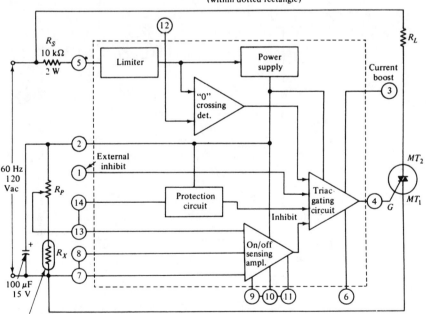

Figure 9.21 Functional block diagram of the CA3059 zero-voltage switch. (*RCA Solid State Division*)

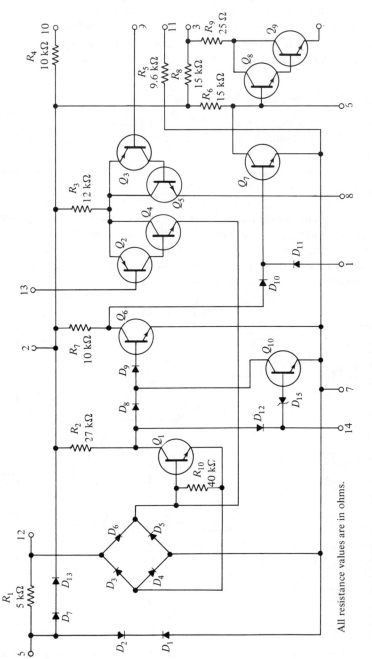

Figure 9.22 Schematic diagram of the CA3059 zero-voltage switch. (*RCA Solid State Division*)

All resistance values are in ohms.

216

4. The *SCR gate driver*, which amplifies the output of the zero-crossing detector sufficiently to supply the moderate current required by the SCR's control-gate input.

5. A *protection circuit*, which senses opens, shorts, and other malfunctions, and removes drive from the SCR input.

As the block diagram of Fig. 9.21 and the internal schematic of Fig. 9.22 illustrate, the internal circuitry of the LIC is never subjected to the large ac line voltage, even though power is derived from the same source, because of an external series-dropping resistor and internal zener-diode clamping. This is another example of the use of a complex, high-gain monolithic circuit to do critical work within the normal LIC process restrictions, while an external element (a triac) inexpensively handles the larger voltage, current, and power dissipation required.[1]

Combination-Control IC: A very interesting and useful control IC is offered in the *RCA monolithic CA3097E device* (Fig. 2.14), *designated as a Thyristor/Transistor Array.* In addition to the thyristor (or SCR) switching element, the 16-pin DIP package includes four other elements: a *PNP/NPN* transistor pair, a programmable unijunction transistor (PUT), a zener diode, and an uncommitted *NPN* transistor. The data sheet for this highly flexible device (*File No. 663*) includes a host of control-application schematics, including a pulse generator, one-shot timer, Schmitt trigger, and series and shunt regulators.

QUESTIONS

9-1. Most OP AMPs require additional power-control elements when the amount of output power exceeds approximately
   (a) 500 mW
   (b) 100 mW
   (c) 25-50 mW

9-2. *Not included* in the class of linear integrated circuits (LICs) are
   (a) Regulators
   (b) Comparators
   (c) Digital interface circuits
   (d) None of these

---

[1] For detailed application information, see the RCA ICAN—4158, "Applications of the CA3059 in Thyristor Circuits."

9-3. The type of IC regulator in which the impedance is continually varied is a
  (a) Linear regulator
  (b) Switching regulator
  (c) Fixed-voltage regulator

9-4. Zener diodes are used in regulators primarily to furnish
  (a) Main regulating action
  (b) Freedom from temperature dependence
  (c) Convenient reference voltage

9-5. The general-purpose 723 type of regulator is designed for
  (a) Dual-polarity (±) output voltage
  (b) Only fixed single-polarity voltage
  (c) Both fixed and adjustable single-polarity voltages

9-6. A regulator type having a minimum external component count, widely used as a local 5-V regulator for digital logic, is
  (a) Type 109
  (b) Type 1568/1468
  (c) Type 3094A

9-7. Zero-voltage switches are used primarily to reduce interference from the following type of power:
  (a) 60-Hz, 115-V ac power
  (b) RF-IF power
  (c) Pulsating dc power

9-8. In the block diagram of a basic voltage regulator (Fig. 9.1), the error between the instantaneous output voltage and the reference voltage is reduced by feedback from the error amplifier which changes
  (a) The base voltage of the control transistor
  (b) The collector of the control transistor
  (c) The emitter of the control transistor

9-9. In the voltage-regulator diagram in Fig. 9.3(a), the operating values are as follows:

Zener (temperature-compensated) voltage (V_z) = 3.9 V
Reference voltage (V_{ref}) = $A_v V_z$, where A_v = 3
Unregulated dc voltage $(V+)$ = 20 to 28 V
Regulated voltage (V_{out}) = $A_v V_z - V_{BE}$, where V_{BE} of the series-pass transistor = 0.7 V)

The final regulated voltage (V_{out}) will be
  (a) 27.3 V
  (b) 19.3 V
  (c) 11.0 V

PROBLEMS

9-1. For a 5-V regulated power supply, a change in the input voltage of 2 V results in a 0.1-V change in the output voltage. What is the percent of input regulation?

9-2. Repeat Problem 9-1 for a 15-V regulated supply.

9-3. Under no load, the output of a regulated supply is 15 V. Under full load, the output falls to 14.95 V. What is the percent of load regulation?

9-4. The no-load voltage of a regulated power supply is 12 V. If the percent of load regulation is 1 percent, what is the full-load voltage?

9-5. The regulated output voltage for the adjustable regulator of Fig. 9.17 is 12 V. Assuming $I_{adj} = 50\ \mu A$, calculate the value of R_2.

9-6. Repeat Problem 9-5 for an output voltage of 5 V.

9-7. Referring to Fig. 9.19, a regulated output of 100 mA is required. Calculate the value of R_1.

9-8. Repeat Problem 9-7 for an output current of 0.5 A.

Chapter 10

Digital-Interface Circuits

10-1 DIFFERENTIAL VOLTAGE COMPARATORS

The integrated-circuit comparator is generally considered under the linear IC category, even though, strictly speaking, it belongs to both the linear and digital worlds. It appears almost universally in digital systems, which require that a logic signal be available only when a dc voltage somewhere in the system is more or less than a *critical threshold value.* It also finds use in many other instances when the level of an analog signal is to be a determining factor.

In its simplest form, it is provided with *two analog inputs* and *delivers a ZERO or ONE output, depending on which input is the larger.* It is thus a two-faced device, which is linear at its input and digital at its output. In this respect it can also be regarded as a basic, one-bit analog-to-digital converter, and so finds use in a wide variety of circuit systems. In addition to interfacing digital circuits, its analog applications include such uses as variable-threshold Schmitt triggers, discriminators, and other level detectors.

The dc voltage comparator is almost as widely used as the OP AMP and, in fact, is very similar to it with respect to the common property of high open-loop amplification. Each of these linear ICs, however, is specifically designed for its particular application, and the differences between them will become more apparent as we examine the properties of the voltage comparator more closely.

The comparator is specifically designed around the properties of *speed* and *accuracy* for differential-comparison purposes. This calls for a deliberate tradeoff of high gain with wide bandwidth (to ensure small propagation delays), and also an output that is compatible with logic levels. Because of the inherently opposing design requirements of high gain versus wide bandwidth (and the consequent tradeoffs between them), there are a number of different types of comparators available. Speed of operation is emphasized by the types designed for wider bandwidths (and consequently smaller propagation delays); other types favor accuracy, which depends on two main factors, one being high gain to reduce the *gain error* (the amount of difference signal required to cause the amplifier to switch), and the other being the minimizing of initial offset error with its attendant temperature drift. These properties are more successfully combined in the specially designed comparator than they are in general-purpose OP AMPS.

10-2 OPERATION OF COMPARATOR

Ideally, a dc comparator consists of an inverting (−) and noninverting (+) input and an output, as shown in symbol form for the basic circuit in Fig. 10.1. Here an unknown voltage (e_{in}) is applied to the inverting terminal 3, while a precise

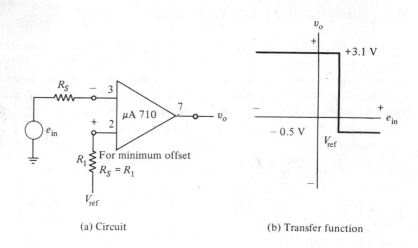

(a) Circuit   (b) Transfer function

Note: Pin nos. apply to metal-can connections only.

Figure 10.1 Basic comparator: (a) circuit in symbol form, compares level of input (e_{in}) with reference voltage (V_{ref}); (b) transfer function, showing ideal change of output from "high" (3.1 V for logic ONE) to "low" (negative 0.5 V for logic ZERO) whenever the input voltage exceeds the reference by as much as 1 or 2 mV. (*Fairchild Camera and Instrument Corporation*)

reference voltage (V_{ref}) is applied to the noninverting terminal 2. During the time that the unknown voltage is less than the reference, the output remains "high" (at a defined logic-compatible level, which is positive 3.1 V in this case). When the unknown voltage becomes only an infinitesimal amount more positive than the reference (1 or 2 mV more positive in this case), the output descends very rapidly to a "low" value (negative 0.5 V), signifying a ZERO, as opposed to the previous value for a ONE. Similarly, when the input connections are reversed (with V_{ref} at the inverting terminal), the output will also be reversed, with a ONE output occurring whenever the unknown input (at the noninverting terminal) slightly exceeds the reference.

10-3 COMPARATOR CHARACTERISTICS

The ideal dc comparator has infinite voltage gain and zero "offset" voltage and current errors at its input, and responds instantly to any change of condition at its input. These are similar to the requirements that apply to ideal OP AMPS; thus the internal construction of most dc comparators is very similar to that of an OP AMP. There are differences, however. The OP AMP must be capable of linearly reproducing an input signal at its output, while this restriction does not apply to the comparator. The OP AMP must be capable of large output voltage swings, while the dc comparator must swing its output between two logic levels suited to the particular logic family it is designed to drive. Certain comparators are designed to produce substantial output currents, so that their "decision" may be used to directly drive a lamp or relay; this output current usually flows in only one direction, whereas high-current OP AMPS must both source and sink large currents. Since the OP AMP is generally used within a dc feedback loop, it is nearly always operating safely within its "linear" range; the dc comparator, on the other hand, is usually overdriven at its input, and must be designed so that it can "recover" quickly from such overdrive and, if possible, do so instantaneously so that very large differences between input voltages do not cause internal damage. Both the OP AMP and the dc comparator must have good common-mode rejection (CMR); that is, they must respond only to the *difference* between input voltages, and not to their individual values. Finally, both types must have high input impedances, negligible input currents, and input errors that are negligibly affected as temperature varies.

The first commercially successful dc comparator is the type *μA710*, shown schematically in Fig. 10.2.[1] In addition to the basic dc comparison and "decision-making" function of the 710 type, the dual versions of this type (type 711) include a *strobing* input, which permits an external logic signal to determine when the output will respond to input differences. The strobe input may be

[1] R. J. Widlar, *Operation of a Fast IC Comparator*, Fairchild Application Bulletin, APP-116.

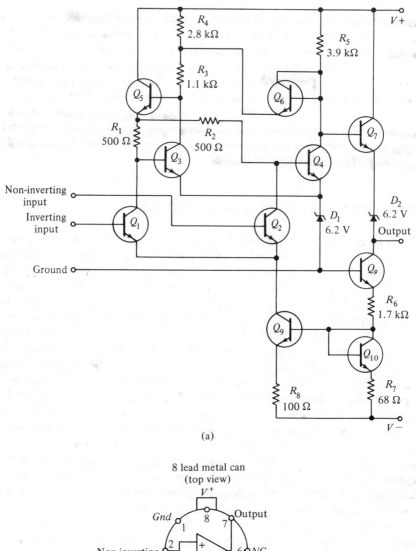

(a)

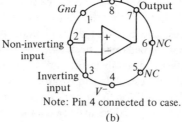

8 lead metal can
(top view)

Note: Pin 4 connected to case.

(b)

Figure 10.2 Schematic circuit of the μA710 comparator: (a) input and reference voltages may be interchanged, depending on desired logic level
at output; (b) connections for metal-can package with V+ = 12 V,
V- = -6 V (also available in 14-lead dual-in-line package). (*Fairchild Camera and Instrument Corporation*)

used, for example, when a system must measure and perform logic based on an input voltage condition at a specific time, but not respond at other times, as in sampled measurement or control systems.

10-4 ADVANCED-PERFORMANCE COMPARATORS

In overcoming the inherently difficult task of producing a comparator that is fast, while simultaneously offering good input characteristics, precision, and versatility, advanced types of comparators have been produced to emphasize particular characteristics in the inherent *tradeoff between speed and precision.*

A limitation in the dual-comparator type *μA711* lies in the fact that while there are two sets of inputs there is only one output terminal (pin 9 in Fig. 10.3). This limitation is overcome in the dual types having two separate outputs, such as *Motorola MC1514*, which also has an improved output capability.

Added versatility to the single comparator type 710 is provided by the ability to operate from a *symmetrical ±15-V supply*, and the *addition of a strobe terminal*, as offered by the *National LM106* type.

Multiple comparators are also available in *quad form*, as, for example, the *National LM339*, the *Motorola MC3302/P*, and *MC3430* series.

High-Speed Types

Other advanced *types emphasizing high speed* include the *Fairchild μA760* (response time of 16 ns) and the *Signetics SE529* (response time of 12 ns).

Precision Types

Among the advanced *precision comparators* is the widely used *National LM111*, considered the workhorse of precision comparators, just as the 710 is for high-speed comparators. (*Note:* the LM111 is a single comparator and not to be confused with the *μA711*, which is a dual-comparator type.)

Although the 111 type is slower than the 710 (e.g., 200-ns response time versus 40 ns), the *input current* I_B (and resulting offset current) is *generally hundreds of times lower* (e.g., 100 nA I_B versus 20 μA).

In the similar category of precision comparators, mention may be made of the *μA734*, offered as useful for analog-to-digital (A/D) conversion use (as discussed in a later section). Similarly, the *Intersil 8001* precision comparator is offered for its low-power consumption (30 mW).

Another precision comparator from *Precision Monolithics*, called *Mono CMP-02*, has a maximum offset specification of only 0.8 mV and typical drift of 1 μV/°C. While its response time of 160 ns is, as expected, greater than the

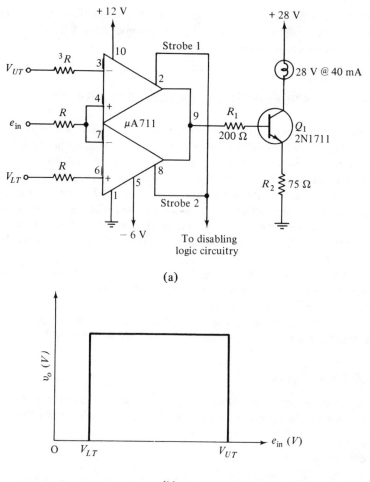

(a)

(b)

Figure 10.3 Dual-comparator (μA711) arranged as a "sense amplifier": (a) shown connected to a lamp-indicator driver, it indicates a change of logic level (ZERO or ONE) to determine when the input (e_{in}) is between two precise reference thresholds (V_{UT} upper test limit and V_{LT} lower test limit); (b) idealized response of sense amplifier. (*Fairchild Camera and Instrument Corporation*)

fast types mentioned above, it is still better than other precision types, which typically have response times of 200 ns or more.

A trend toward including additional circuitry on the comparator chip is exemplified by the μA750, which features a self-contained voltage-reference capability.

10-5 SENSE AMPLIFIERS

While the dc comparator gives a logic output when an unknown input is above or below a reference voltage, it is often desirable to determine when an unknown input is *between* two precise reference thresholds. This can be accomplished by using two dc comparators, one for each threshold, and performing appropriate logic on their outputs so that the resultant logic output is ONE if the unknown is between the thresholds, and ZERO if outside the thresholds. Such a dual-threshold comparator is called a *sense amplifier.*

Sense amplifiers are LICs that are used in the heart of most digital computers. Magnetic memories in such computers, whether cores, plated wires, rotating drums, or magnetic tape, all give very small output signals (in the millivolt range) when interrogated. The interrogation process, in the case of core memories, requires a large current to pass through *address* wires running through the core in question; the few millivolts of response picked up in another wire are in the presence of a much larger, noisy "spike," resulting from the interrogation. The sense amplifier has good common-mode rejection, allowing it to respond differentially to the small memory content signal while ignoring the much larger interrogation signal. The sense amplifier is also strobed so that it transmits information to the computer only at times when such information is available and needed.

A typical sense-amplifier application of the $\mu A711$ is shown in Fig. 10.3(a). It essentially consists of two type 710s on one chip, plus common output logic circuitry. Figure 10.3(b) shows the idealized characteristics of a sense amplifier. An output level only exists when the input signal is between V_{LT} and V_{HT} volts. Other typical dual sense amplifiers are the *SN7524* and *SN7424* types from *Texas Instruments.*

Five-Step Analog Level Detector

Texas Instruments TL489C level detector (Fig. 10.4) consists of five comparators and a reference voltage network for detecting an analog input signal. Output Q_1 is switched to a low logic level at a typical input voltage of 200 mV. After each 200-mV step, subsequent outputs are switched to low logic levels. All outputs are switched to low logic levels at a typical input voltage of 1 V. The open-collector outputs are capable of sinking currents up to 40 mA and may be operated at voltages up to 10 V. The analog input has a typical impedance of 100 kΩ.

Because all five trigger points have a switching hysteresis of typically 10 mV, the circuit may be operated with slow input signals without the danger of oscillation at the outputs. To prevent pickup of noise, a capacitor should be connected between the high-impedance input and ground, especially when the input is driven from a high-impedance source.

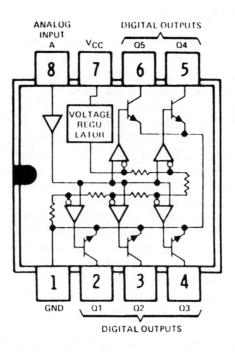

ANALOG
INPUT
A   Vcc   DIGITAL OUTPUTS
          Q5   Q4

VOLTAGE
REGU
LATOR

1   2   3   4
GND Q1   Q2   Q3

DIGITAL OUTPUTS

Figure 10.4 The TL489C five-step analog level detector. (*Texas Instruments*)

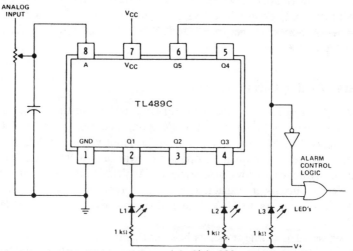

ANALOG
INPUT

Vcc

8   7   6   5
A   Vcc Q5   Q4

TL489C

GND Q1   Q2   Q3
1   2   3   4

ALARM
CONTROL
LOGIC

L1   L2   L3   LED's

1 kΩ   1 kΩ 1 kΩ
                     V+

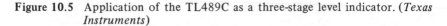

Lamp L1 is turned on at input voltages (pin 8) ≥ 200 mV and the alarm turns off.
Lamp L2 is turned on at input voltages ≥ 600 mV to indicate correct operation.
Lamp L3 is turned on at input voltages ≥ 1000 mV and the over-range alarm turns on.

Figure 10.5 Application of the TL489C as a three-stage level indicator. (*Texas Instruments*)

The unit is especially designed to detect and indicate analog signal levels. It may be used in various applications such as low-precision meters, warning signal indicators, A/D converters, feedback regulators, pulse shapers, delay elements, and automatic range switching. The power outputs are suitable for driving a variety of display elements such as LEDs or filament lamps. The outputs may also drive digital integrated logic such as TTL and CMOS. Its application as a three-stage level indicator is illustrated in Fig. 10.5.

10-6 ANALOG-TO-DIGITAL AND DIGITAL-TO-ANALOG CONVERTERS

Physical quantities are usually continuously varying (analog) signals. Examples of such varying quantities include pressure and temperature. Before these signals can be processed by a digital system, such as a microcomputer, they are converted to digital form. This is the purpose of the analog-to-digital (A/D) converter, or ADC. Because the output of a digital system is in binary form, it may be required to convert the binary signal to an analog signal. This is accomplished by the digital-to-analog (D/A) converter, or DAC.

D/A Converters

A binary number is a string of 1's and 0's. For example, binary number 1101 is equal to

$$1 \times 2^3 + 1 \times 2^2 + 0 \times 2^1 + 1 \times 2^0 = 13_{10}$$

where subscript 10 indicates that the number is expressed in the base 10 (decimal) number system. Any binary number, N, can be represented in the following form:

$$N = a_n 2^n + a_{n-1} 2^{n-1} + \cdots + a_1 2^1 + a_0 2^0$$

where coefficients a_n, a_{n-1}, etc., are either a 0 or a 1. This is the kind of signal a D/A converter converts to an analog equivalent.

An example of a practical D/A converter, in simplified form, is illustrated in Fig. 10.6. An R-$2R$ ladder network is connected to the inverting input terminal of an OP AMP. Each end of the resistor whose value is $2R$ is returned to the arm of a SPDT switch. (The replacement of the mechanical switch with an electronic one will be considered later.) The switch at the farmost right represents the most significant bit (MSB); the switch at the farmost left represents the least significant bit (LSB). If a bit is a 0, the switch is returned to ground; if a 1, it is returned to reference voltage V_R. The output voltage, V_a, is expressed by

$$V_a = - V_R \left(\frac{a_n}{2} + \frac{a_{n-1}}{4} + \cdots + \frac{a_1}{2_n} + \frac{a_0}{2^{n+1}} \right)$$

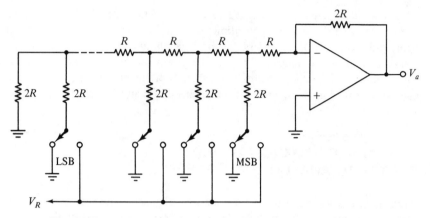

Figure 10.6 Basic structure of a R-$2R$ ladder D/A converter.

Assume that $N = 1101 = 13_{10}$ and $V_R = -16$ V; then

$$V_a = 16\left(\frac{1}{2} + \frac{1}{4} + \frac{0}{8} + \frac{1}{16}\right) = 13 \text{ V}$$

If a varying analog input signal is used instead of the fixed reference voltage V_R, the device is called a *multiplying D/A converter*.

D/A Switches

Numerous electronic switching circuits are available to replace the mechanical switches in Fig. 10.6. One example of such a switch whose signal is derived from a CMOS gate is shown in Fig. 10.7. The output of the gate drives a voltage follower. This driving circuit does not load down the R-$2R$ network, thereby providing an accurate output of the converter.

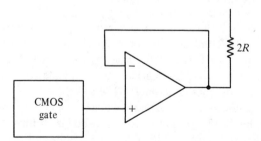

Figure 10.7 Basic electronic switch employing a CMOS gate and a voltage follower.

Commercial D/A Converters

There are many manufacturers producing D/A converters and, in this section, representative examples of these devices are examined. Currently available D/A converters have resolutions of 8, 10, 12, 14, and 16 bits. An example of a 12-bit unit is *Hybrid Systems CAC356* series, whose block diagram is provided in Fig. 10.8. It dissipates low power, typically 75 mW, and is compatible with DTL, TTL, and CMOS logic.

An example of a multiplying D/A converter is the *MC1508* series. Referring to the block diagram of Fig. 10.9, the device consists of a reference current amplifier, an *R-2R* ladder, and eight high-speed current switches. For many applications, only a reference resistor and reference voltage need be added. The switches are noninverting in operation; hence, a high state on the input turns on the specified output current component. The switch uses current steering for high speed and a termination amplifier consisting of an active load gain stage with unity gain feedback. The termination amplifier holds the parasitic capacitance of the ladder at a constant voltage during switching, and provides a low impedance termination of equal voltages for all legs of the ladder.

A/D Converters

The three most commonly used A/D converters are the *successive-approximation, dual-slope,* and the *parallel,* or *flash,* types. The most popular is the successive-approximation A/D converter. It exhibits high conversion speed and excellent accuracy and stability. The dual-slope converter is used often in digital panel meters and in multimeters. Although slower than the successive-approximation type, the dual-slope A/D converter has excellent accuracy and stability. The parallel-type converter has the fastest conversion time. Because of the large number of comparators used in this type of converter, it is difficult to achieve high resolution.

Successive-Approximation A/D Converter

A block diagram of a successive-approximation A/D converter is illustrated in Fig. 10.10. This type of converter operates by comparing the analog input signal to a series of trial conversions. The first trial compares the input to the value of the MSB, or half of full scale. Figure 10.11 shows the progression of trials for a 3-bit converter. If the input is greater than the MSB value, the MSB is retained and the converter moves on to try the next most significant bit, or three-quarters full scale. If the input had been less than the MSB, the logic would have turned the MSB off before going on to the next most significant bit, or one-quarter full scale. This branching continues until each successively smaller bit has been tried, with the entire process requiring $n + 1$ trials.

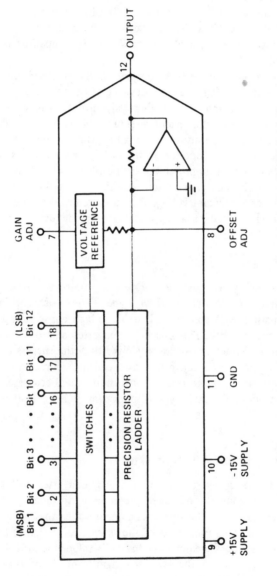

Figure 10.8 A functional block diagram of a 12-bit DAC. (*Hybrid Systems Corporation*)

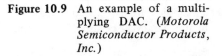

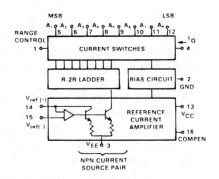

Figure 10.9 An example of a multiplying DAC. (*Motorola Semiconductor Products, Inc.*)

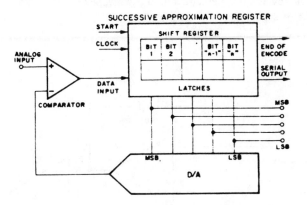

Figure 10.10 Functional block diagram of a successive-approximation ADC.

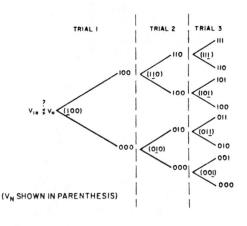

Figure 10.11 Flow diagram for a 3-bit successive-approximation analog-to-digital conversion.

(V_N SHOWN IN PARENTHESIS)

Referring to Fig. 10.10, a start command inserts a 1 in the MSB position of the shift register. This first latch is also set to 1. If the output of the comparator is *greater* than the analog input, the latch is reset to zero. If, however, the output of the comparator is *less* than the analog input, the 1 remains in the MSB position of the latch, and the MSB position of the counter is a 1. At the next clock pulse, a 1 is placed in the bit 2 position of the shift register. This sequence of events is repeated until all the bits have been tried. After the last bit has been processed, the "end of encode" changes state, indicating that the conversion has been completed and the parallel output digital data may be used. The output also is available in serial form.

An example of a commercial successive-approximation A/D converter is the *Analog Devices AD571* 10-bit converter. A monolithic device, full-scale calibration accuracy is ±0.3 percent and the conversion time is 25 μs. No external components are required for a conversion.

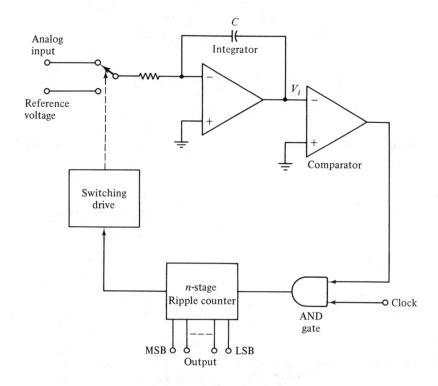

Figure 10.12 Block diagram of a dual-slope A/D converter.

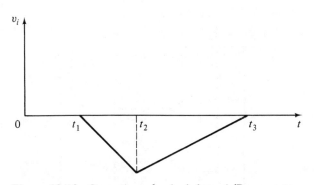

Figure 10.13 Operation of a dual-slope A/D converter.

Dual-Slope A/D Converter

A block diagram of a dual-slope A/D converter is provided in Fig. 10.12. Assume that the analog input voltage is greater than zero (i.e., it is positive) and the reference voltage is less than zero (it is negative); this is referred to as *unipolar operation*. Referring to Fig. 10.13, at time t_1, the analog voltage is connected to the input of the integrator and is integrated for a fixed number of clock pulses. The n-stage ripple counter automatically resets itself to zero at time t_2, and the input switch connects the reference voltage to the input of the integrator. Because the reference voltage is negative, the output of the integrator is positive going. As long as v_i is negative, the output of the comparator is positive, thereby enabling the AND gate. At time t_3, the AND gate is inhibited and no further clock pulses are counted. The output of the counter, during the time interval t_2-t_3, is proportional to the analog input signal. An example of a ramp type A/D converter is the *Motorola MC14443* and *14447* series.

Parallel A/D Converter

Referring to Fig. 10.14, several comparators are used with an ascending order of reference voltages in the parallel ADC. For an n-bit conversion, $2^n - 1$ comparators are required. Each comparator is biased by a string of resistors connected to the reference voltage. The analog signal is connected to the noninverting inputs of the comparators, which are all connected together.

Those comparators having references below the instantaneous value of the input signal will give 1 outputs, while the others will give 0 outputs. The resultant logic outputs can be fed into additional logic processors and transmitted. The accuracy of such a system depends on how many comparators are used, the spacing of their reference voltages, and how often the logic informa-

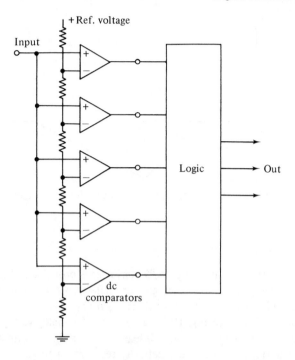

Figure 10.14 Parallel A/D converter: a series of comparators with an ascending set of reference-voltages to supply the corresponding ONE and ZERO outputs that make up the digital-logic output.

tion is sampled and transmitted. The $\mu A 734$ comparator may be used in parallel A/D converters with 12-bit accuracies and a 1-megabit conversion rate.

10-7 BORDERLINE LINEAR-DIGITAL INTEGRATED CIRCUITS

As may be seen from the numerous type numbers mentioned for comparators, sense amplifiers, and A/D and D/A converters, there is a healthy competition for these kinds of LIC digital-interface circuits. This rivalry results in a multiplicity of types for the digital-interface units that approaches the amount of proliferation of the digital ICs for use in digital computers. Accordingly, within the limits of scope and size here, further specification details of these digital-interface circuits (*including line drivers and line receivers*) are best left to texts and manuals that concentrate on digital ICs.

10-8 COMPLEMENTARY MOS (CMOS) TYPES

A good example of the overlapping trend between the linear and digital types is furnished by the complementary MOS types. In the digital field these FET types have made great inroads with their simplified circuitry and low power consumption, especially in switching applications. An extension to the LIC field of linear-circuit processing is offered by the *CD4007A*, an RCA COS/MOS IC, which provides a dual complementary pair, plus inverter, as the D/A switch and OP-AMP output stage for a digital-to-analog converter.[2]

In addition to the switching applications of the CMOS (complementary MOS) type, there are many instances where it can be used beneficially as an alternate to the *linear bipolar IC*. Since it is possible to arrange the CMOS circuit to perform *linear amplification* (within its smaller voltage-amplification limits), it may often be a preferred choice in special cases where its low power-consumption feature is an important factor.[3]

QUESTIONS

10-1. The comparator type of LIC differs from the OP-AMP in the following characteristic:
   (a) Speed and accuracy
   (b) Linear reproduction of input-signal
   (c) Generally retaining a dual (±) power supply

10-2. The strobe input of a comparator is used
   (a) To determine the level of response
   (b) To determine the time of response
   (c) To determine the polarity of response

10-3. Comparator LICs are generally operated
   (a) With no feedback
   (b) With large amounts of feedback
   (c) With small amounts of feedback

10-4. A sense amplifier responds to
   (a) A single polarity
   (b) A single threshold level
   (c) A dual threshold level

---

[2] "D/A Conversion, Using the CD4007A (COS/MOS IC)," RCA Application Note, ICAN-6080.

[3] See RCA brochure, "Linear Applications of COS/MOS."

10-5. For a constant dc input to an integrating OP-AMP, the output waveform is
   (a) A constant dc output
   (b) A rising ramp output
   (c) A sawtooth output

10-6. The advantage of dual-slope integration in A/D conversion in a DVM is
   (a) A double-amplitude digital output
   (b) The digital output requires the clock frequency to be only half its nominal count
   (c) The digital output does not depend on the precise accuracy of the clock frequency

10-7. Complementary MOS (CMOS) circuits have certain outstanding design advantages for
   (a) Switching circuits
   (b) Linear circuits
   (c) Both switching and linear circuits

PROBLEMS

10-1. Referring to Fig. 10.6, determine the output if $N = 1011$. Assume $V_R = -16$ V.

10-2. Repeat Problem 10-1 for $N = 1111$.

10-3. If, in Fig. 10.6, $V_R = -10$ V and $N = 1101$, what is V_a?

10-4. Repeat Problem 10-3 for $N = 1111$.

Chapter 11

Precision and Instrumentation Operational Amplifiers

11-1 MAJOR HIGH-PERFORMANCE CHARACTERISTICS

Beyond the group of general-purpose OP AMPS previously discussed (Chapters 4 and 5), there is a large (and growing) list of advanced types—each one offering improved specifications of one kind or another. These represent the continuing advances in the innovative design and fabrication techniques of IC technology, along with the trend by designers toward the steady adoption of improved LIC devices in place of many complicated circuit functions that were formerly feasible only in discrete-circuit form.

The improved devices provide *high-performance properties that are primarily aimed at overcoming potential error-producing characteristics of LICs* in three main areas, as follows:

1. *Decreasing input bias currents* (I_B) to obtain a higher input-impedance (Z_{in}) value (as with FET input), and to reduce errors associated with such input currents flowing through large resistors.
2. *Reducing drifts* generally associated with the temperature coefficients (TC) that cause changes in initially trimmable offsets, such as *input-voltage offset* $[V_{io}(TC)$ in $\mu V/{}^{\circ}C]$ and *input-current offset* $[I_{io}(TC)$ in $pA/{}^{\circ}C]$
3. *Reducing inherent noise voltages* that are large enough to be objectionable in the amplification of low-level signals.

There are other improved specifications that are particularly desirable for certain applications (such as requirements for *wide bandwidth* or for *micropower dissipation*), and these requirements will be considered separately as they appear in such applications.

11-2 DEVELOPMENT OF PRECISION OPERATIONAL AMPLIFIERS

Considering the continuing appearance of newer OP AMPs with ever-better refinements, it might be thought that such high-performance amplifiers are a fairly recent trend. Not so; these efforts at precision amplifiers go all the way back to the vacuum-tube era, when such OP AMPs were satisfying a keen demand for improving the accuracy of the then-emerging *analog computers*. Similar efforts were subsequently applied to developing the discrete-transistor versions of these precision amplifiers to replace the bulky tube affairs with their mechanical choppers.

The present activity centers around the vastly more reliable and efficient compact devices produced by IC techniques, with heavy accent on the ICs inherently better matching of differential-transistor pairs, with accompanying improvement, for example, in common-mode rejection capability.

Superbeta Transistors in OP AMPS

The use of the superbeta (or supergain) transistors is an outstanding example of an intensive development in the improvement of OP AMPs. This has resulted in a substantial reduction in the input bias current (I_B) by about three orders of magnitude, with a corresponding improvement in input impedance (Z_{in}) or (R_{in}). The exceedingly small bias currents used are shown in the characteristic curves of Fig. 11.1(a), along with a functional schematic of the supergain voltage-follower operation in part (b).[1]

By means of advanced processing in IC diffusion technology, the superbeta transistor can be fabricated with a *current gain exceeding 4,000*, but with its breakdown voltage at only about 4 V. However, when the IC chip is subjected to two separate emitter diffusions, it is possible to build the superbeta and the standard *NPN* transistors on the same chip. By this means (not available in discrete circuits), such IC circuits are designed to take advantage of high current gain, and yet can be operated at high voltages.

The superbeta-transistor types of OP AMP (exemplified by the National LM108/A and Motorola MC1556) have input currents reduced to less than one

---

[1] R. C. Dobkin, "New Developments in Monolithic Op Amps," *Electronics World*, July 1970.

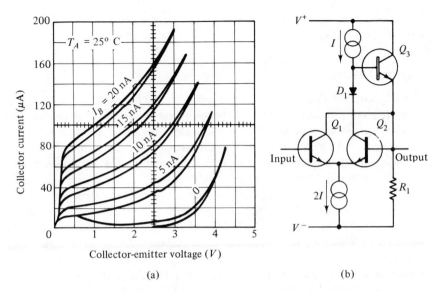

(a)   (b)

Figure 11.1 The superbeta (or supergain) transistor: (a) super-β transistor output curves; (b) functional diagram of voltage-follower circuit. Super-β transistors Q_1 and Q_2 are operated at very low bias currents (in nanoamperes), while buffer transistor Q_3 is the only one operated at relatively high voltage.

tenth of the ordinary types, and offer a design alternative to the FET-input type (such as the Fairchild μA740). These examples offer high input-impedance alternatives that overcome the FET limitations (particularly with respect to the bothersome leakage current of the junction FET, which doubles for every 10°C rise, as discussed later).

11-3 SELECTING OPERATIONAL AMPLIFIERS FOR HIGH-PERFORMANCE CHARACTERISTICS

It is well to note at this point that the emphasis in this chapter on small error-producing factors is not generally a major concern in a multitude of amplification applications, where the OP AMP is sufficiently close to an "ideal amplifier"; this emphasis, on the other hand, is primarily important in cases where *precise analog computation or highly accurate signal-conditioning applications* are involved. It is in these areas that efforts at technical refinements are constantly being made to overcome inherent device limitations, such as temperature dependence, dc offset drift (not wholly susceptible to elimination by negative feedback), and ever-present noise, however small. This constant attention results in a wide spectrum of high-performance OP AMPs with a variety of specially tailored specifications.

Figure of Merit

If some standard figure of merit for a precision amplifier could ever be agreed upon, it would undoubtedly be a very handy aid in the selection process. But considering the facts of life, it is unreasonable to expect that a single figure should accurately evaluate the greatly different performance characteristics of amplifiers which are designed to emphasize particular properties. However, with this caution in mind, one such attempt to furnish a not too unreasonable comparison of overall performance may be mentioned.[2] This *comparison* includes a formula for arriving at an arbitrary figure of merit by including in the numerator such figures as voltage gain (V/V) and slew rate (V/μs) and dividing the numerator product by the product of input-voltage offset (V_{io}), voltage-offset drift (μV/$^\circ$C), bias current (I_B), and supply current (I_{CC}). In practice, this notion of a single figure of merit value would be valid only if proper weights for the pertinent parameters were assigned by each user for particular requirements.

Chart of Typical High-Performance OP-AMP Characteristics

For practical initial selection purposes, we may divide those refined characteristics that are desired for a particular OP AMP application into two main categories, given in Table 11-1, for quick comparison with general-purpose specifications. The main error categories are as follows:

    1. Input bias current (I_B) and accompanying input impedance (Z_{in}).
    2. *Input-voltage offset drift* (V_{io} in μV/$^\circ$C).

    (*Note: Noise voltage* specifications are most important in instrumentation amplifiers, and are discussed in Section 11-9.)
    Each main category is discussed in a separate section that follows, along with a representative example (or examples) of each type.[3]

---

[2] Based on a performance-comparison chart compiled by J. Talley (of Texas Instruments): using the relative values described above, his figure of merit places the TI superbeta type SN52771 at 600, between the extremes of 0.4 (for the 741) up to a maximum beyond 1,000 (for the 108).

[3] For particular applications, very helpful guides to OP-AMP selection may be found in the product catalogs of many manufacturers, particularly in those with concentration on the precision-instrumentation types of LICs, such as Analog Devices, Teledyne/Philbrick, and Burr–Brown. (See Appendix IV for a more complete list of LIC manufacturers and their addresses.)

TABLE 11-1
Typical Precision OP-AMP Characteristic Values

	Low-Cost Types				Premium Types		
	General-Purpose Type	Precision Type	Superbeta Types		FET Input	Chopperless	High Speed (BW to 100 MHz)
	μA741	μA777	LM108A	MC1566	AD503	AD517	AD507
Input bias current (I_B) (nA)	≈100	25	2	15	0.01	2	2
Input impedance (Z_{in}) (MΩ)	>1	>2	>30	5	to 10^{12} Ω	4	40
Drift of offset voltage (V_{io} TC) (μV/°C)	—	15	<5	<15	25	0.5	15
Slew rate (V/μs)	0.5	0.5	0.2	2.5	6	0.1	35

For micropower types (where values are adjustable), see Section 11-9.
For instrumentation type (AD520) (which is monolithic, but not strictly an OP AMP), see Sections 11-9 and 11-10.

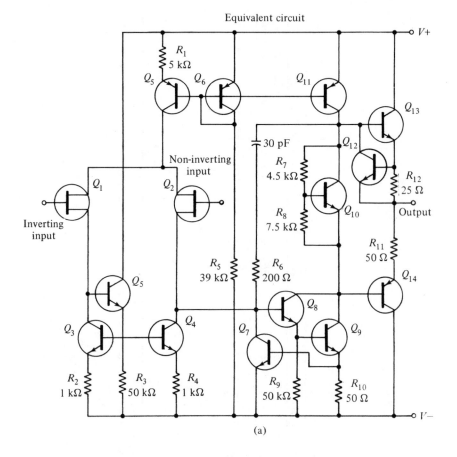

(a)

(b)

NOTE: Pin nos. apply to 8-pin package (either
8-pin metal can or 8-pin mini-DIP types).

Figure 11.2   FET-input OP AMP: (a) schematic diagram of μA740 equivalent circuit, showing the FET differential pair in the input circuit; (b) pin connections are same as internally compensated 741 type. (*Fairchild μA740 Data Sheet*)

11-4 LOW-BIAS-CURRENT (I_B)—HIGH-INPUT-IMPEDANCE (Z_{in}) TYPES

The value of input bias current (I_B) for this group of OP AMPs (and modular LICs) is given in nanoamperes, indicating a reduction by a factor of 10 or more compared to general-purpose OP AMPs. This reduction is achieved by designs using either superbeta transistors (as discussed in a previous section) or FET types in the input circuit (see μA740 in Fig. 11.2). A comparison with the general-purpose OP AMPs of the 741 type can more easily be seen at a glance by the listing in Table 11-2 of *representative rough values for comparable types.*

TABLE 11-2
Comparison of Input Impedance (Z_{in})

Type	I_B (nA)	Z_{in}
General-purpose type 741	Approx. 100	Approx. 2 MΩ
Superbeta type MC1556	Approx. 10	Approx. 5 MΩ
Superbeta type LM108A	Approx. 1	Approx. 50 MΩ
FET-input type μA740	Approx. 0.1	Up to 10^{12} Ω
FET-input type AD503	Approx. 0.01	Up to 10^{12} Ω

The value for low bias current is naturally increasingly important in dealing with *high source resistances*, and also for improved accuracy in *integrators* and in *sample-and-hold circuits.*

Similarly, accuracy is improved by a low value of I_B to reduce input-current offsets. Here the *ultrahigh Z_{in} of the FET-input types*, while essential for some very high input-impedance sources, *may not be preferable to the superbeta transistors*, because the FET has the more rapid drift with temperature, as discussed in the next section.

11-5 LOW VOLTAGE-DRIFT TYPES

Where an application requires extreme stability against voltage drifts with time or temperature, the traditional method has been to use *chopper-stabilized amplifiers*, where the "low-frequency" components most subject to drift have been chopped (or modulated), and are then processed in a practically driftless ac-coupled amplifier, which also handles the higher frequencies. This method is incorporated in *chopper-stabilized* OP AMPs, such as *Analog Devices type 234*, to achieve voltage drifts as low as 0.1 μV/$^\circ$C and averaging about 2 μV/month. These are generally available in *modular*, rather than monolithic, form, but there are noteworthy exceptions, such as *TI's SN62/72088*, which is in the 14-pin DIP form. Another, in standard TO-99 form, is the *Harris model HA2900*, which can also serve as a good example of the improved drift performance achieved in the chopper-stabilized models. Compared to the standard 741 types, the specifications for this chopper-stabilized model of OP AMP offer better than an order-of-magnitude improvement in its low value for offset-voltage drift of 0.2 $\mu V/°C$, a figure that is associated with an extremely small input-offset current (I_{io} = 0.05 nA) and an exceptionally large open-loop gain (A_{VOL} = 5 × 10^8 or greater than 170 dB).

Chopperless low-drift types have been designed in monolithic form, and these retain the advantage of the differential OP AMP with its high common-mode rejection ratio (CMRR)—an advantage that must be sacrificed in using the single-ended chopper-stabilized amplifier. Such chopperless types as the *AD517*, a superbeta example, still offer a low value of voltage drift of 0.5 $\mu V/°C$ and an input bias current (I_B) around 1 to 2 nA.

While the value of I_B for the superbeta type *AD517* results in an input impedance of only a few megohms (compared, say, with that offered by the FET-input type *AD 503* of up to 10^{12} Ω), it becomes important to realize the limitation imposed by the FET on its current-drift specification, which doubles for every 10°C rise in temperature. Hence, from the complete low-drift standpoint, the superbeta transistor may emerge as superior for its certified low-drift specification for a monolithic type.

Typical applications for the low-drift type of OP AMP include *accurate summing amplifiers* (as in servo loops), *precision comparison and regulation*, and generally as stable input amplifiers of low-level signals, where voltage drifts would become a significant part of the input signal, or would otherwise degrade the accuracy of operation.

Combination of Low-Drift and FET-Input Properties

An example of a high-performance monolithic amplifier that combines low input voltage-offset drift with FET-input performance is offered by the *Burr–Brown model BB3521* series. The overall performance in this type (and in similar high-performance types from other manufacturers) is achieved by advanced fabrication techniques, such as *active laser trimming*. The features include the expected high input impedance (from 10^{11} Ω up) of the input FET pair. This monolithic pair is carefully matched to keep the initial input offset voltage down to a small value (250 μV) and also to hold its drift down to a low level of 1 $\mu V/°C$.

When considering the ultrasmall input bias current (I_B) in FET types (10 pA in this case), we must also keep in mind the accompanying FET characteristic that calls for this input current to double for every 10°C rise. However, where temperature variations are acceptably low, this combination of the desirable properties of high input impedance and low voltage drift allows the use of the OP AMP for high-accuracy requirements (such as 0.01 percent buffer use), while still retaining a configuration similar to a general-purpose 741 type.

11-6 ADDITIONAL HIGH-PERFORMANCE FEATURES

The limited listing of typical high-performance specifications in Table 11-1 was confined, for purposes of clarity, to just the few major characteristics of low input bias current (I_B) with corresponding high Z_{in}, and low voltage-offset drifts ($\mu V/°C$), as being two salient properties in characterizing precision types of OP AMPS. In addition to these, other high-performance features are desirable for particular purposes; such extra factors give rise to special types, such as the following:

1. *Wide-band (or high-speed) types* (discussed in Section 11-7).
2. *Low-power (or micropower) types* (discussed in Section 11-8).
3. *Instrumentation (or low-noise and high CMRR) types*, discussed in Section 11-9, summarizing most high-performance features.

A discussion of these additional types by no means exhausts the full range of advanced types in the active development of OP AMPS (and other LICs); it does, however, within space limitations, offer a sound basis for viewing the diverse performance capabilities of OP AMPS (including and beyond the general-purpose types) and should, hopefully, enable an intelligent interpretation of the maze of technical specifications given in manufacturers' data sheets. As a further aid, examples of these types are identified in Appendix II, in addition to a *cross-referenced* list of manufacturers' types and model numbers in Appendix III.

11-7 WIDE-BAND (HIGH-SLEW-RATE) TYPES

The property of wide frequency response is allied with high speed (i.e., fast rise and fall times) by the constant relation of their product; this is generally given in terms of the -3-dB high-frequency cutoff (f_c) of the amplifier and rise time (t_r) of a square wave output as a constant product:

$$(f_c)\,(t_r) = 0.35$$

Unity-Gain Bandwidth

Consider first the bandwidth of an OP AMP: this extends from direct current to high-frequency cutoff (f_c). This ability to amplify a wide band of frequencies is generally specified in terms of the *-3-dB bandwidth of the* OP AMP at unity gain. This is typically the same value as the *gain–bandwidth product* ($G \times BW$), as exemplified by the value of 1 MHz for the general-purpose 741 type. Here the -3-dB bandwidth is 10 kHz at a gain of 100, giving a $G \times BW$ product of 1 MHz (100×10 kHz); similarly, when used with a greater amount of feedback

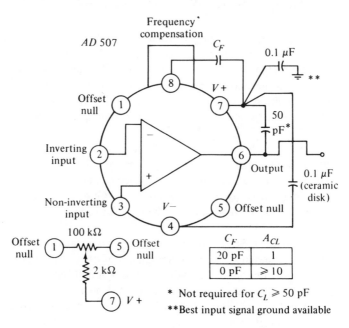

C_F	A_{CL}
20 pF	1
0 pF	$\geqslant 10$

* Not required for $C_L \geqslant 50$ pF
**Best input signal ground available

(a)

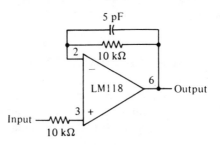

Fast voltage follower

(b)

NOTE: Pin nos. apply to 8-pin package (either
8-pin metal can or 8-pin mini-DIP types).

Figure 11.3 Wide-band and high-speed monolithic OP AMPs: (a) wide-band
(100 MHz) type AD507 (*Analog Devices*) connection diagram; (b)
high-speed voltage-follower, with feed-forward compensation
circuit of the LM118 type (*National Semiconductor*) to boost
slew rate beyond 100 V/μs.

for unity gain (at which gain it is still unconditionally stable because of its internal compensation), the unity-gain bandwidth is given as 1 MHz (1 X 1 MHz). Also, in the general-purpose class, the *extended bandwidth* 748 type is capable of extending the bandwidth at a gain of 100 to a frequency of 100 kHz (by the use of a 3-pF external compensation capacitor), giving a $G \times BW$ product 10 times greater to 10 MHz (100 X 100 kHz). However, for cases requiring still larger bandwidths, we must turn to the high-performance wide-bandwidth types, such as, for example, the *AD507*, which provides a gain–bandwidth product out to 100 MHz in an OP AMP configuration, as shown in Fig. 11.3(a). It will be seen in the figure that some external compensation (20 pF at terminal 8) is required to go down to unity gain, although this is not required for closed-loop gains greater than 10.

Slew Rate

When a large-amplitude signal is fed into an OP AMP, the full output appears after the lapse of a small amount of time, owing to the presence, generally, of internal capacitance. This limitation on the *rate at which the output can change (or slew)* in response to an input step is expressed as the *slew rate in volts per microsecond* (V/μs). Since the same limiting factor of capacitance (internal or distributed) is involved in both bandwidth and slew rate, it can be expected that the high-speed amplifier types will also be found in the wide-band group.

In the case of the example of the wide-band amplifier (type *AD507*), the slew rate is specified as 35 V/μs, as compared to the value of 0.5 V/μs for the general-purpose 741 type, an improvement of 70 times.

This improvement in bandwidth and slew rate is obtained, as can be expected, by various tradeoffs of other high-performance characteristics. When compared, for example, with the precision superbeta type *AD517*, which is optimized for low-voltage-drift properties, the offset-voltage-drift is only 0.5 μV/°C, contrasted with a drift of around 15 μV/°C for the *AD507*.

With regard to the *correspondence between wide bandwidth and slew rate*, it should be noted that the correspondence is not strictly a constant, depending on the particular fabrication techniques employed. For example, the precision high-speed *LM118* type offers a small-signal bandwidth of only 15 MHz (much below the 100-MHz figure of the *AD507*), while still providing a guaranteed slew rate of 50 V/μs (as opposed to 35 V/μs for the *AD507*). Moreover, in the *LM118*, which is internally compensated down to unity gain, it is possible to use external feed-forward compensation for inverting applications that will boost the slew rate to over 150 V/μs, and almost double the bandwidth. A fast voltage-follower circuit for the *LM118* is shown in Fig. 11.3(b), where a feed-forward compensation of only 5 pF is required for obtaining a generally ample high-speed action; similarly, a simple capacitor can be added to reduce the *0.1 percent settling time to under 1 μs*. Used in such high-speed applications as oscil-

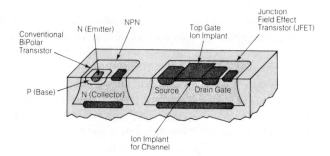

Figure 11.4 Bi-FET process employing ion implantation of gate and channel. (*National Semiconductor*)

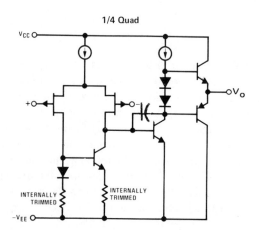

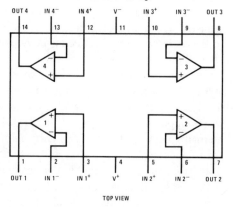

Figure 11.5 The LF347 quad OP AMP: (a) simplified schematic; (b) connection diagram. (*National Semiconductor*)

lators, active filters, and sample-and-hold circuits, the *LM118* represents an order of magnitude better performance than the original wide-band $\mu A 709$ type, a standard long used in industry as the first popular OP AMP. A similar improvement over the original $\mu A 709$ type is seen in the $\mu A 715$, which offers a bipolar combination that can be frequency compensated for unity-gain bandwidth of 65 MHz, and a slew rate of 100 V/μs. The *Burr-Brown 3554* OP AMP provides a slewing rate of 1,000 V/μs.

An example of a wide-bandwidth quad JFET OP AMP is the *National LF347*. It uses an improved Bi-FET process that is based on ion implant technology (Fig. 11.4) to reduce the input offset voltage and noise. Laser trimming of the resistors in the input stage reduces the input offset voltage to 2 mV or less. The bandwidth is 4 MHz and the slew rate is 13 V/μs. A simplified schematic and a connection diagram are provided in Fig. 11.5. The OP AMP also is available as a single unit (*LF351*) and as a dual (*LF353/354*).

11-8 MICROPOWER TYPES

Low-drain performance characteristics for *battery-powered applications* are provided by the micropower types of OP AMPS. A representative example, *National type LH0001A*, offers a *no-load power dissipation of less than* $\frac{1}{2}$ mW at supply voltages of ±5 V. With this small amount of dc input power, there is still an ample open-loop gain of more than 30,000 (or around 90–dB), with a typical low noise-voltage value of 3 μV rms. These specifications allow for quite satisfactory operation in many low-power instrument and transducer amplifiers. A premium version of the micropower OP AMP, the *CA3078A* (from *RCA*), is tailored for operation from a single supply, using a 1.5-V battery (size AA cell). Operation in this fashion, as an inverting 20-dB (gain of 10) amplifier, is shown in Fig. 11.6; the total power consumption for this circuit (and also for the noninverting version) is approximately 675 nW (or less than $\frac{3}{4}$ μW). The output voltage swing is 300 mV (or about 107 mV rms).

The operating point of the *CA3078A* in the circuit of Fig. 11.6 is obtained by an external resistor (R_{set} = 30 MΩ), connected from the positive supply to terminal 5. Other operating points may be realized by adjusting R_{set} to change the microamperes of quiescent current (I_Q) over a wide range of 3 orders of magnitude (from 1 to 1,000 μA). This provides for a resulting change in open-loop gain (A_{VOL}) of roughly from about 60– to 90–dB (or from about 1,000 to 30,000 times). Such amplifiers, having the ability of changing bias current (and so gain or transconductance) by an externally adjusted resistor, have been designated by RCA as *operational transconductance amplifiers* (OTA).

In the OTA category there is type *CA3080A* where the amplifier-bias input terminal allows variation of the amplifier bias current (I_{ABC}) from 0.1 to 1,000 μA, to produce corresponding changes in the transconductance (g_m)

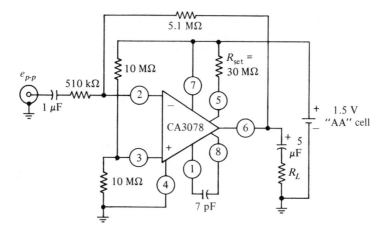

Figure 11.6 Micropower OP AMP circuit: the CA3078 type, operating with a single supply of 1.5 V, consumes less than 1 μW of quiescent power; in this 20-dB inverting amplifier circuit, the bias-setting resistor (R_{set}) is externally adjustable. (*RCA Solid-State Division Data Book SSD-201*)

from 2 up to 20,000 μS. Although not strictly a micropower amplifier, the device power dissipation, at a supply voltage of ±3 V, can be made to change from just over 1 to 10,000 μW. A similar scheme for controlling micropower operation is called the *programmable* OP AMP, in *type μA776*. The quiescent current is again controlled by an external resistor (R_{set}). By selecting values for this resistor to produce an I_{set} current variation, the input bias current can be changed from about 0.1 to 20 nA. As a result, with a supply voltage of ±3 V and I_{set} = 1.5 μA, the resulting supply current of 15 μA produces an input power of 90 μW (15 μA X 6 V). With R_{set} = 22 MΩ and a power supply of ±1.2 V, the power dissipation (P_d) for a 20-dB gain is reduced to 600 nW (or 0.6 μW). A low-cost version of the μA776 is *Motorola's MC3476*. Its power supply voltages range from ±6 to ±18 V.

11-9 INSTRUMENTATION AMPLIFIER

The amplifier requirements for instrumentation purposes are among the most severe for any group of linear ICs. In addition to all the properties needed for precision amplifiers (as previously discussed), there is also the added consideration of handling the changing conditions produced by small transducer variations on accurate operation. This requires the instrumentation amplifier to preserve its initial high accuracy against expected variations of its input sensor or transducer while rejecting any undesired voltages. The strategies for meeting these requirements can also serve as a practical summary of the refined OP AMP characteristics that are significant in applications demanding very stable and highly accurate amplifiers.

The situation can be clarified by examining the instrumentation amplifier in its frequent use for amplifying the output of a transducer-bridge configuration,[4] shown in functional-equivalent form (as if it were a single OP AMP) in Fig. 11.7. Here the signal of interest (e_d) is the difference between the voltages at the corners of the bridge $(e_1 - e_2)$, as a result of one of the four resistors, acting as a sensor, changing its resistance. The output signal (e_d)

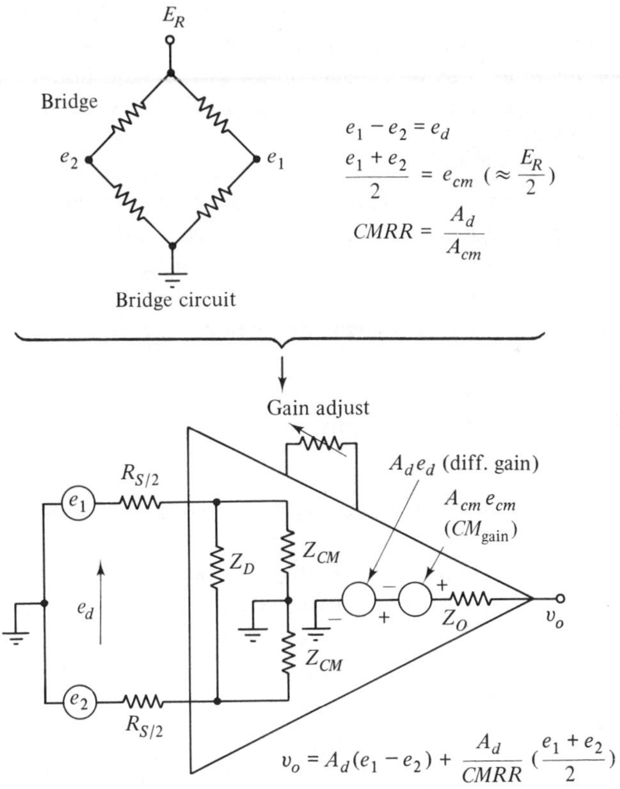

Figure 11.7 Functional diagram of instrumentation amplifier; the equivalent input from the balanced bridge at the left is shown connected as a differential input to the amplifier equivalent at the right. (*Analog Devices*)

[4] Analog Devices: *Product Guide.*

contains a superimposed common-mode voltage (e_{cm}) roughly equal to half the supply voltage ($E_R/2$). As the sensor changes its resistance ($R + \Delta R$), the bridge output becomes the difference signal (e_d), which is given in both its exact and approximate forms as the following[5]:

$$\text{Exact:} \quad e_d = E_R \frac{\Delta R}{4R + 2\Delta R};$$

$$\text{Approx. (for small deviations):} \quad e_d = E_R \frac{\Delta R}{4R},$$

where $2\Delta R \ll 4R$.

The difference signal (e_d) is applied as a double-ended input to the operational amplifier, by using both inputs of the amplifier, which then acts as a *differential amplifier*; this type of amplifier should (and initially does) have a satisfactorily high common-mode rejection ratio (CMRR), usually well beyond 80–dB (or 10,000:1).

As shown in the left-hand portion of Fig. 11.7, each input originally has an equal source resistance ($R_s/2$), under which condition the output (v_0) contains a practically negligible amount of amplified common-mode interference. Before examining what happens as the bridge is unbalanced, it is instructive to see how this differential configuration of the amplifier accomplishes the highly desirable rejection of common-mode interference.

Advantages of Differential Input

The beauty of the differential input of an op amp lies in the desirable feature of amplifying a small difference between two large, but nearly equal, signals. This is accomplished by *symmetrical circuitry*, obtained by using both the inverting (−) and noninverting (+) inputs of the amplifier, as shown in Fig. 11.8. A good insight into the working of this symmetrical circuit[6] can be gained by regarding it as a combination of a unity-gain inverter [with e_1 applied to the inverting (−) terminal] and a voltage follower [where e_2 is applied to the non-

[5] The derivations of the unbalanced-bridge output are given in S. Prensky, *Electronic Instrumentation*, 2nd ed., Prentice-Hall, Inc., Englewood Cliffs, N.J., 1971.

[6] A delightful analysis of this circuit is given (even including humor) in J. I. Smith, *Modern Operational Circuit Design*, John Wiley & Sons, Inc. (Interscience Division), New York, 1971, Chapter 5. This circuit and its variations are also well analyzed in G. E. Tobey, L. P. Huelsman, and G. G. Graeme, *Operational Amplifiers*, McGraw-Hill Book Company, New York, 1971.

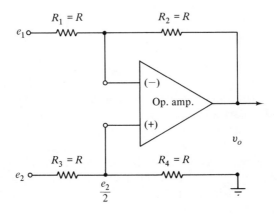

$$v_o = K(e_2 - e_1),$$

where $K = \dfrac{R_2}{R_1} = \dfrac{R_4}{R_3}$ [= 1, when all resistors = R]

Figure 11.8 Symmetrical arrangement of instrumentation amplifier, using differential (double-ended) input connection; the output (v_o) is an accurate representation of only the difference between the input voltages ($e_2 - e_1$) with excellent, common-mode rejection ratio (CMRR).

inverting (+) terminal]. When all four resistors have the same value (R), the output of the circuit (v_o) is unity times the input difference:

$$v_o = 1(e_2 - e_1), \quad \text{with generally negligible } e_{cm}.$$

An analysis of the circuit by superposition is instructive to show how the symmetrical circuit produces an output directly proportional to the input difference. With all four resistors equal to R, we have the following two conditions for superposition:

First, for $e_2 = 0$, the output is for an inverting amplifier, which equals

$$- e_1 \frac{R}{R} = -e_1.$$

Second, for $e_1 = 0$, the output is for a noninverting voltage follower with a gain of 2 [or $1 + (R/R)$], acting on $e_2/2$ (or half the input signal), making this output equal to

$$2 \frac{e_2}{2} = +e_2.$$

Then, by superposition, adding the two outputs,

$$v_o \text{ (overall)} = e_2 - e_1.$$

The action of the common-mode voltage on the symmetrical circuit is shown on the right side of Fig. 11.7 to simulate the same undesired signal (e_1 + e_2)/2 on both inputs. Under this condition, the high value of CMRR allows the common-mode voltage to swing within fairly wide limits, while the output magnifies only the difference $e_1 - e_2$ by the differential gain (A_d). Thus, even if the differential gain is only unity, the symmetrical circuit still has a great advantage over a single-ended configuration; in rejecting the unwanted common-mode voltages, it retains the important property of high CMRR. Moreover, this concept is very useful in *computation circuits* serving as *subtracter* and also as *adder-subtracter* functions.

While the symmetrical concept is highly attractive in theory, the simple circuit of Fig. 11.8 *does not work out* to be equally good for a practical instrumentation amplifier. The necessity of working with four accurately matched resistors makes the calibration for a given gain quite difficult. For a gain (K) greater than 1, it is necessary to keep resistance ratios carefully matched to satisfy the relation

$$\frac{R_2}{R_1} = \frac{R_4}{R_3}$$

Bridge Amplifiers

An additional problem presents itself when the input voltages e_1 and e_2 are derived from an unbalanced bridge. This arises from the relation previously given for the output of a bridge which is unbalanced by a resistance change in the active transducer that forms one arm of the bridge, $R + \Delta R$. This output is linear for only small fractional deviations ($\Delta R/R = \delta$), as seen in the formulas expressed in terms of

$$\text{Exact } v_o = \frac{E_R}{2} \frac{\delta}{2 + \delta},$$

$$\text{Approx. } v_o \approx \frac{V}{4} \delta.$$

As a rule of thumb, for a given deviation, say δ = 5 percent, the discrepancy in the linear output given by the approximate formula is about half (or error = 2.5 percent).

A circuit that overcomes this limitation on wide deviations is shown in

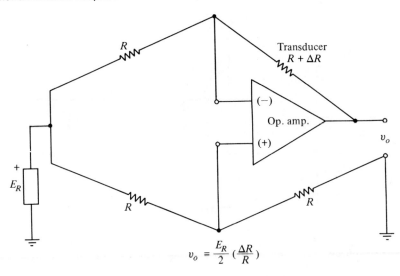

$$v_o = \frac{E_R}{2}\left(\frac{\Delta R}{R}\right)$$

Figure 11.9 *Linear* bridge output circuit, where the differential input connection to the OP AMP produces a bridge output directly proportional to the fractional deviation ($\delta = \Delta R/R$) of the transducer element ($R + \Delta R$).

Fig. 11.9 where the output remains linear with the deviation δ. This circuit[7] is drawn to show its relation to the traditional bridge, and its output, a *linear readout* of the resistance deviation δ, is given as

$$v_o = \frac{E_R}{2}\delta,$$

which is an exact form, as compared to the preceding similar-looking formula, which is only approximate for the traditional bridge.

Another trouble with the simple version of the symmetric difference amplifier of Fig. 11.8 is that it *lacks the high-resistance inputs* that are desirable in an instrumentation amplifier. One possible method of overcoming this limitation would be the use of two additional OP AMPS to serve as high-input-impedance voltage followers feeding the third OP AMP connected as a difference amplifier. This basic idea is shown in Fig. 11.10(a), with the resistance ratio for gain (K) obtained by the ratio KR/R.

This theoretical method still leaves us with the difficulty of tailoring the gain (preferably variable) by the awkward need for changing the exactly matching ratios of four transistors, with the added chances of changing the CMRR value as the gain is changed.

[7] Smith, *Modern Operational Circuit Design.*

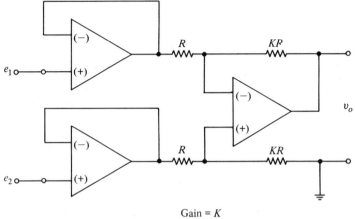

Gain = K

(a)

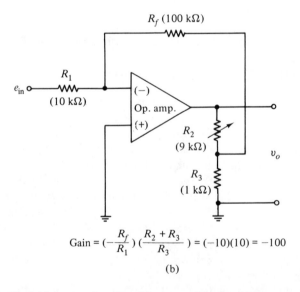

$$\text{Gain} = \left(-\frac{R_f}{R_1}\right)\left(\frac{R_2 + R_3}{R_3}\right) = (-10)(10) = -100$$

(b)

Figure 11.10 Variable gain circuits: (a) symmetrical arrangement for an instrumentation amplifier versus (b) an inverting OP AMP circuit; the use of three amplifiers in (a) is one method of preserving high Z_{in} and CMRR, despite changes in gain.

Variable Gain

The methods for obtaining variable gain in an instrumentation amplifier are necessarily different from the simple circuits that are employed with a general-purpose OP AMP. For general use, the gain can ordinarily be changed by simply changing the feedback resistor (R_f) to change the gain-determining ratio R_f/R_1. However, it is well to note that this easy method has a severe limitation in some cases, owing to the fact that the larger values of R_f produce correspondingly larger values of voltage offset, caused by the increased drop $(I_B R_f)$ as R_f increases. This becomes more evident in dealing with a circuit where the signal source, for example, requires a fairly large input resistor R_1, of, say, 10 kΩ, and we wish to increase the gain to 100. This would require making R_f equal to 1 MΩ, which might easily cause an unacceptably high voltage offset.

A useful method for avoiding undesirably high values of R_f is shown in Fig. 11.10(b), where the feedback connection for R_f is taken from a voltage divider across the output. In the example, where we have set R_1 equal to 10 kΩ, we can use a lower value of 100 kΩ for R_f and still obtain a gain of 100 by connecting to the tap on the voltage divider, which supplies only one tenth the output to R_f. In this case, the gain is obtained by the R_f/R_1 ratio of 10 multiplied by the reciprocal of the voltage-dividing ratio $(R_2 + R_3)/R_3$, also 10, to give the desired gain of 100.

This method illustrates a *general rule of thumb that a voltage-dividing network in the feedback path produces the opposite effect of such a network in the forward path.* This relation can be intuitively verified by regarding a smaller feedback voltage as equivalent to a larger resistor in determining the amount of feedback current.

With regard to the instrumentation amplifier, we find that neither of the preceding gain variations is feasible because of the necessity of maintaining accurate resistance ratios to preserve the ability of the amplifier to reject common-mode voltages.

Practical Approach to Instrumentation Amplifier

A different approach[8] to accomplish practically all the desired objectives for an instrumentation amplifier, and to largely overcome the difficulty with variable gain, requires the use of only two OP AMPs, as shown in Fig. 11.11(a). It permits the gain variation to be set by an external GAIN ADJUST (R_3) separate from the accurately matched resistors making up R_1 and R_2 that are required for the symmetric arrangement. This circuit possesses the very desirable feature that the CMRR value is completely independent of the setting of the gain adjust R_3. As

[8] Smith, *Modern Operational Circuit Design.*

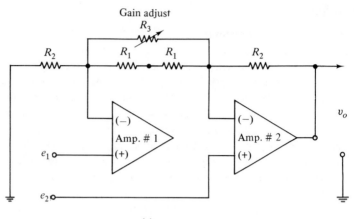

(a)

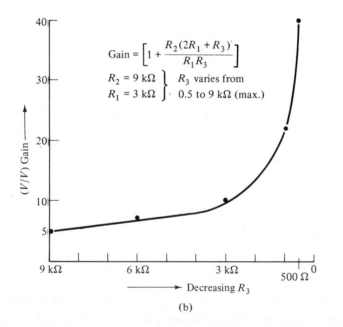

(b)

Figure 11.11 Condensed version of instrumentation amplifier circuit, using only two amplifiers: (a) the gain adjustment of external resistor R_3 does not affect the accuracy or CMRR, which depends on close-matching of external resistors R_1 and R_2; (b) gain variation.

a result, with the use of precision ratios for R_2/R_1, the gain can be set and easily changed without degrading the amplifier performance.

The gain formula, along with the graph of increased gain as the resistance of R_3 is changed to lower values, is shown in the plot of Fig. 11.11(b). The curve is seen to be quite nonlinear. In the example, using a 9-kΩ rheostat at its full value for gain adjust R_3, the gain is seen to increase quite regularly as the resistance is lowered for a gain increase of 5 to 10, but there is a sharp non-linear increase (up to a gain of 40) as the resistance is further decreased to a minimum value of 500 Ω. Accordingly, to obtain a wide range of gain variation, a tapered (or vernier type) of variable resistor would be called for.

It can be seen from the previous discussion that many strict characteristics are required for an instrumentation amplifier, and it should therefore be no surprise that the resulting form of this type of amplifier would tend to be more advanced than a single OP AMP even of the precision variety. This, in fact, is the proud claim of many of the instrumentation amplifiers that have appeared in *module form*. It is also true in the case of the *monolithic form* of the representative example discussed in the next section, where the manufacturer's initial description emphasizes the fact that it is not an OP AMP, and that, although it can be made to operate in a similar fashion, it has measurable properties beyond those found in any single OP AMP.

11-10 REPRESENTATIVE INSTRUMENTATION AMPLIFIER (MONOLITHIC)

In monolithic integrated-circuit form, *Analog Devices AD520* represents an instrumentation amplifier that includes two OP AMPs and associated circuitry in its internal closed-loop gain block, all in a 14-lead DIP package. The circuit configuration is shown in Fig. 11.12(a), accompanied by the pin configuration in part (b). Here it will be noted that provision is made for connecting an external resistor (R_{gain}) and another for R_{scale}, with the closed-loop gain thus determined for any desired value from 1 to 1,000 by the ratio

$$A_{VCL} = \frac{R_{scale}}{R_{gain}}.$$

A most important result for this arrangement is the fact that the *gain adjustment does not degrade either the input impedance (Z_{in}) or the rejection ability* (CMRR), two critical factors in an instrumentation amplifier. Thus the *AD520* performs like a modular (or hybrid) instrumentation amplifier, even though its monolithic form gives it the appearance of an OP AMP.

As an instrumentation amplifier, it should not be confused with the class of OP AMPs that also provides the feature of variable gain setting by an external

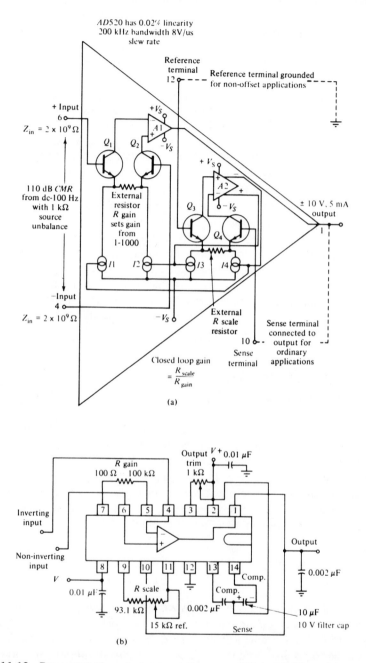

(a)

(b)

Figure 11.12 Representative instrumentation amplifier AD520: (a) circuit configuration shows the two internal amplifiers with provision for external resistors R_{gain} and R_{scale}; (b) pin configuration of the 14-pin DIP, showing additional terminals for SENSE and REF connections used in the applications of Fig. 11.13. (*Analog Devices*)

resistor (such as the operational transconductance amplifier mentioned previously). As examples of the difference, in the *AD520*, both inputs are at high input impedance (2,000 MΩ), and it also bears repeating that *the Z_{in} and CMRR remain high at all gain settings.* (The obstacles in the way of achieving these properties with any single OP AMP were described in the previous section.)

Additional features, also shown in Fig. 11.12, include the optional use of the REF and SENSE terminals. The *output reference terminal* (REF) allows the output offset to be adjusted independently of gain setting, and the high-impedance *sense terminal* (SENSE) allows the circuit's feedback to be derived from either the output terminal or an arbitrary external point.[9]

Using the REFERENCE Terminal

An application where the amplifier output is connected to a recorder is shown in Fig. 11.13(a). In many cases the input (e_{in}) is from a transducer bridge, and it is

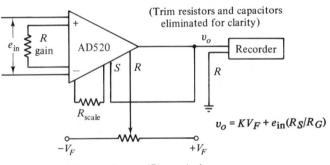

$$v_O = KV_F + e_{in}(R_S/R_G)$$

(a) Using reference (R) terminal

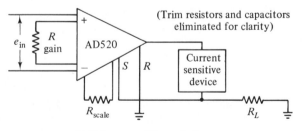

(b) Using sense (S) terminal

Figure 11.13 Applications using AD520 optional terminals: (a) use of REF (R) terminal to position recorder pen; (b) use of SENSE (S) terminal to stabilize current through a floating component, such as a solenoid. (*Analog Devices*)

[9] These and other instrumentation considerations are comprehensively discussed in the eight-page data sheet of the Analog Devices AD520 J/K/S (see Appendix IV for addresses).

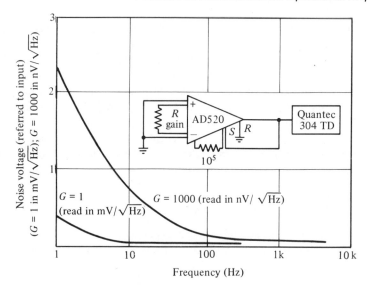

Figure 11.14   Noise voltage graph for instrumentation amplifier gains of 1 and 1,000. (*Analog Devices*)

desirable to position the pen of the recorder at a desired (or set point) part of the chart; this is accomplished by connecting the arm of an external voltage divider to the REF terminal. (Ordinarily, the REF terminal is grounded.)

Using the SENSE Terminal

In ordinary use, the SENSE terminal is connected to the output, and it thus provides the internal feedback link. In Fig. 11.13(b), this terminal is used to sense a voltage across a given load that is driven through a current-sensitive device. The current in this floating device, such as a sensitive relay or solenoid, is kept equal to the v_o/R_L, where v_o, as before, is $e_{in}(R_{scale}/R_{gain})$.

Noise

Low noise is obviously an important property in examining specifications for instrument amplifiers; it is generally given in the data sheet in the form of a graph as noise voltage (referred to input), as $mV/\sqrt{Hz}$, over a range of frequencies. This noise voltage is shown for the *AD520* in Fig. 11.14, reading in millivolts for unity gain and in nanovolts for a gain of 1,000.

11-11 MODULAR FORM OF INSTRUMENTATION AMPLIFIER

Most instrumentation amplifiers are packaged in modules of varying complexity. A representative modular example, *Teledyne/Philbrick model 4253*, is shown in

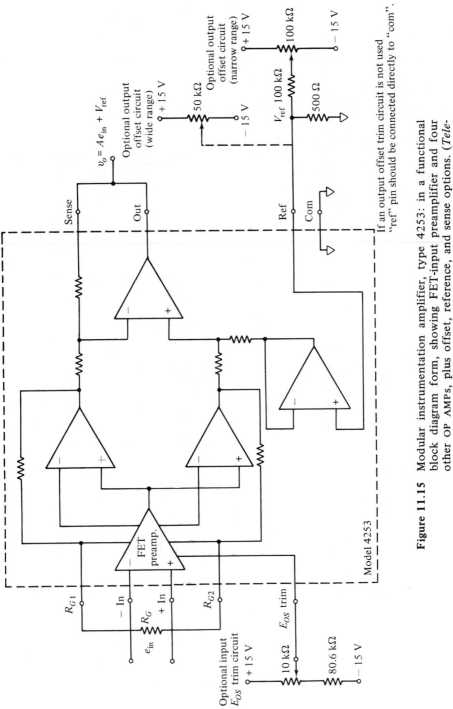

Figure 11.15 Modular instrumentation amplifier, type 4253: in a functional block diagram form, showing FET-input preamplifier and four other OP AMPS, plus offset, reference, and sense options. (*Teledyne/Philbrick*)

265

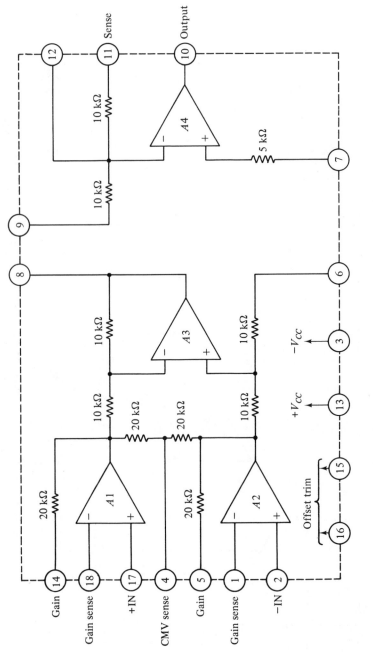

Figure 11.16 Simplified schematic of the 3630 instrumentation amplifier. (*Burr–Brown*)

block form in Fig. 11.15 for a low-drift FET instrumentation amplifier. The functional block diagram shows the use of five internal amplifiers,(including an FET preamp), with the use of external gain resistor R_G providing an adjustable gain of 1 to 5,000, according to the formula

$$\text{gain} = 1 + \frac{100 \text{ k}\Omega}{R_G}.$$

Trim circuits for the reference and sense terminals (discussed in the previous section) are also shown.

The *Burr–Brown 3630* instrumentation amplifier, shown in simplified form in Fig. 11.16, is an example of a thin-film hybrid design. The input stage ($A1$ and $A2$) consists of premium-type bipolar OP AMPs. Connected in a non-inverting configuration, they provide an input resistance of 10^{10} Ω. The second stage ($A3$) is an OP AMP connected as a unity-gain difference amplifier. The four 10-kΩ resistors are matched to ensure that a minimum CMRR of 106–dB at 60 Hz for gains greater than 100 is maintained over temperature and time.

The third stage ($A4$) allows such functions as implementation of an active low-pass filter and additional gain. All the resistors are part of a single thin-film network of nichrome deposited on a passivated silicon substrate. Critical resistors, like the four 10-kΩ resistors used with $A3$, are laser trimmed.

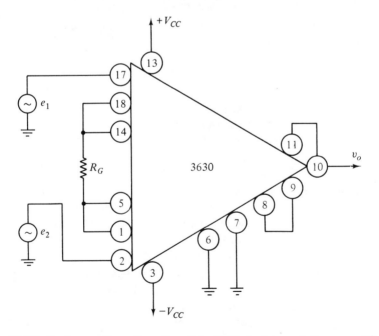

Figure 11.17 Basic configuration for the 3630 instrumentation amplifier. (*Burr–Brown*)

The basic configuration for the 3630 is shown in Fig. 11.17. External resistor, R_G, determines the gain of the amplifier:

$$\text{gain} = 1 + \frac{40 \text{ k}\Omega}{R_G}.$$

For good gain accuracy, R_G should have a low temperature resistance coefficient.

QUESTIONS

11-1. The action of superbeta transistors in reducing the bias currents for instrumentation amplifiers is characterized by
   (a) Darlington connection of the transistors in the IC chip
   (b) Complementary connection of the transistors in the IC chip
   (c) Large current gains at small collector voltages

11-2. The main disadvantage in the use of superbeta transistors is
   (a) Small voltage gain
   (b) Small breakdown voltage
   (c) Small collector current

11-3. Micropower types of OP AMPs can be operated by power-supply voltages as small as
   (a) ± microvolts
   (b) ±1.5 V
   (c) ±5 V

11-4. A good instrumentation LIC must combine the following characteristics in a single package:
   (a) High common-mode rejection with a high fixed gain
   (b) High common-mode rejection with high voltage supply
   (c) High common-mode rejection with variable gain

11-5. The disadvantage with using a single OP AMP in a differential connection in Fig. 11.8, for use as an instrumentation amplifier, is
   (a) Maintaining exact resistor ratios for variable gain
   (b) Maintaining a high gain
   (c) Maintaining a high input impedance

11-6. The use of a bridge-circuit for obtaining the output of a transducer (or sensor) is preferred because the bridge provides
   (a) Greater sensitivity
   (b) Zero output for zero input
   (c) Greater input impedance

11-7. In a precision instrumentation amplifier, the use of three OP AMPs, as in Fig. 11.10(a), instead of a single one, as in Fig. 11.10(b), allows changes in gain while preserving
  (a) High Z_{in}
  (b) High CMRR
  (c) Both high Z_{in} and high CMRR

PROBLEMS

11-1. A square wave is applied to an amplifier. If the rise time of the wave at the output should not be less than 1 μs, determine the cutoff frequency of the amplifier.

11-2. Referring to Fig. 11.7, assume that $A_d = -10,000$, CMRR = 120-dB, $e_1 = 20$ mV, and $e_2 = 10$ mV. Calculate the output voltage, v_o.

11-3. Repeat Problem 11-2 for a CMRR = 80 dB.

11-4. Referring to Fig. 11.10, determine the ratio of R_2/R_3 for a gain of -150.

11-5. Repeat Problem 11-4 for a gain of -250.

11-6. Referring to Fig. 11.11(b), calculate the gain for $R_3 = 500$ Ω, 3 kΩ, 6 kΩ, and 9 kΩ. Compare your results with the values given in the graph.

11-7. For the 4253 instrumentation amplifier, calculate the value of R_G for a gain of 1 and 5,000.

11-8. For the Burr-Brown 3630 instrumentation amplifier, calculate the value of R_G for a gain of 5 and 300.

Chapter 12

Specialized Linear-Integrated-Circuit Applications

12-1 EXPANSION OF LINEAR-INTEGRATED-CIRCUIT FUNCTIONS

In the developments that continue to take place in both the design and the fabrication of linear ICs, we find newer capabilities appearing beyond the more or less "standard" functions. The previous chapters have indicated improvements that have been made in recent years in providing higher input impedances (as in the FET-input types), greater output-current ratings, and greatly refined specifications for improving the accuracy performance of instrumentation amplifiers.

A host of advanced application designs has been keeping pace with the device improvements, producing a large variety of new ways of using LICs. This chapter presents a sampling of selected specialized applications. In some cases the availability of these OP AMPs produces a simpler and more effective circuit, as in *active filters, electrometers*, and techniques for *low noise* in amplifiers of low-level signals. In other instances we find that the "linear" IC can be very profitably employed in nonlinear applications, such as *logarithmic amplifiers* and *function generators*. In addition to such selected circuits, a system application is presented in this chapter for a *digital voltmeter* (DVM), which emphasizes the effective use of linear ICs as building blocks, thus serving to greatly simplify the design of an entire system.

One advanced type, an OP AMP *that can be programmed to fit a specific*

271

application, offers an important concept in extending the flexibility of the OP AMP as a building block. Examples of the use of such programmable OP AMPs are presented in the next section.

12-2 PROGRAMMABLE AMPLIFIERS

This group of OP AMPs offers amplifiers whose characteristics can be altered over a wide range by the selection of an external resistor to set the bias current. This feature was previously introduced in Section 11-8 and in the circuit of Fig. 11.6, where the *CA3078* and *CA3080* were given as examples of an *operational transconductance amplifier* (OTA). This is a term used by RCA to indicate the capability of varying the transconductance (g_m) from 2 to 20,000 μS by changing the amplifier bias current (I_{ABC}) through the selection of the desired external resistance.

In the Fairchild model $\mu A776$, this external resistor is called R_{set} for producing a low bias current (I_{set}), thus allowing *micropower operation* from two battery cells (± 1.2 V), for specialized battery applications, as shown in Fig. 12.1(a).

This flexibility of operation is also extended to *high-output-current applications* by the *programmable amplifier-power switch*, model *CA3094/A* (this type was previously mentioned in Section 9-8 as a power-control-switch device). Its use here as a *linear power amplifier* is shown in Fig. 12.1(b). While itself capable of delivering 3-W average power, it is shown in this circuit as a driver stage for a 12-W amplifier circuit feeding a complementary output stage of power transistors. The programming pin (terminal 5) permits varied special applications such as strobing, squelching, controlling gain (AGC), and mixing for amplitude modulation.

12-3 OPERATIONAL AMPLIFIER AS A REGULATED VOLTAGE SOURCE

A simple and effective voltage source for producing a variable regulated-voltage output, using a simple zener diode, is shown in Fig. 12.2. This circuit makes use of a basic OP-AMP property, where the negative feedback acts to keep a constant voltage difference between the two input terminals. The current from the V+ supply through resistor R_Z establishes the zener-regulated voltage as a reference voltage (V_{ref}), and this is applied to the noninverting (+) terminal of the OP AMP. The application of negative feedback from the output-voltage divider determines the ratio R_2/R_1, and the output voltage (v_o) is

$$v_o = V_{ref}\left(1 + \frac{R_2}{R_1}\right).$$

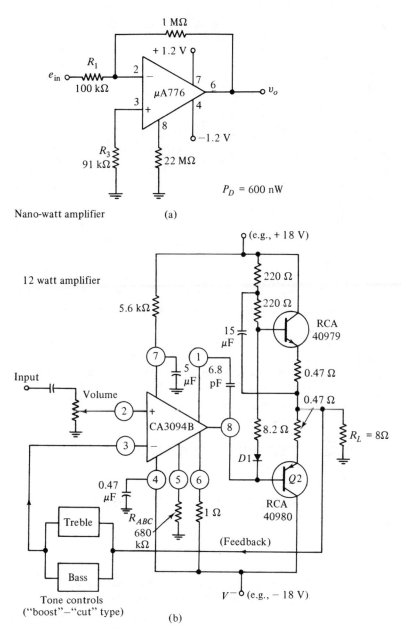

Nano-watt amplifier

(a)

12 watt amplifier

(b)

NOTE: Pin nos. apply to 8-pin package (either
8-pin metal can or 8-pin mini-DIP types).

Figure 12.1 Programmable IC amplifiers: (a) nanowatt amplifier using the
μA776 (*Fairchild Camera and Instrument Corporation*), with setting resistor $(R_{set}) = 22$ MΩ; (b) programmable amplifier/power
switch CA 3094 (*RCA Solid State Division*) in a 12-W configuration with external transistors, and amplifier-bias-current resistor
$(R_{ABC}) = 680$ kΩ.

273

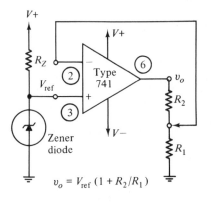

Figure 12.2 Voltage-source circuit: the general purpose OP AMP provides a voltage source which may be varied by the output-voltage divider and which is regulated by the zener diode.

$$v_o = V_{ref} (1 + R_2/R_1)$$

NOTE: Pin nos. apply to 8-pin package (either 8-pin metal can or 8-pin mini-DIP types).

For example, using a 5-V zener and a voltage divider of 15 kΩ tapped at 5 kΩ from ground, the regulated output voltage at v_o would equal 5 V (1 + 2), or 15 V.

When operated within the current limitations of the OP AMP, even a general-purpose type (such as the 741) will produce a very convenient voltage source. A larger output-current capability can be obtained by the use of a booster amplifier following the 741 (as in Fig. 7.1) or with the use of other dc power amplifiers described in Chapter 7.

12-4 CURRENT-SOURCE APPLICATION

A current source, Fig. 12.3(a), can conveniently be realized by inserting the load (R_L) in the feedback path of the OP AMP. For a given input voltage (e_{in}), the current in the load is given simply as

$$I_L = \frac{e_{in}}{R}.$$

A precision current-source application[1] is illustrated in Fig. 12.3(b). Here the output of the OP AMP feeds an FET (as a source follower) to drive a bipolar output transistor and produces the output current I_0. A Darlington connection can be used in place of the FET bipolar combination in cases where the output current is high and the small base current of the Darlington input would not cause a significant error. For negative e_{in}, the output current (I_0) is again

$$I_0 = \frac{e_{in}}{R_1}.$$

[1] *Linear Applications*, a manual by National Semiconductor (see Appendix IV for addresses).

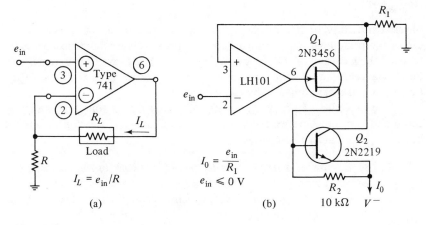

$I_L = e_{in}/R$

$$I_0 = \frac{e_{in}}{R_1}$$

$e_{in} \leqslant 0$ V

(a)　　　　　　(b)

NOTE: Pin nos. apply to 8-pin package (either
8-pin metal can or 8-pin mini-DIP types).

Figure 12.3   Current-source circuits: (a) simple configuration for constant-current (I_L) through load; (b) precision current source, using LH101 (*National Semiconductor*) and external transistors (see text).

The OP AMP used here, the *LH101*, is internally compensated for unity gain (as in the 741 type). If extended-bandwidth types are used (such as the LM101/A or μA748, they would employ an external compensation capacitor for this purpose (usually 30 pF for the types mentioned).

The impedance of this precision current source is essentially infinite for small currents, and is accurate for the usual cases where e_{in} is much greater than the offset voltage, and I_0 is much greater than the bias current (I_B).

12-5   ACTIVE FILTER APPLICATIONS

In classical filter theory, the *passive components* of resistance, inductance, and capacitance (RLC) are traditionally used in network analysis for theoretical explanation and laboratory verification. This practice is educationally sound in forming clear concepts of frequency-selective circuits; it gives a good picture of how the increasing impedance with frequency of inductor L is effective in low-pass filters, along with the opposite action (decreasing impedance of capacitors C, as employed in high-pass filters). These factors, together with the value of Q (the ratio of reactance to resistance, which determines the sharpness of tuning), all form a solid theoretical basis for analyzing filter circuits, whether of low-, high-, or band-pass types.

In practice, however, it is preferable to use active RC filter circuits that make use of amplifiers, especially for low frequencies, where the large size of the inductance becomes awkward and often impractical.

With the convenient IC as the amplifier in an active filter circuit, the physical difficulty of fabricating coils on a chip remains as an obstacle to the use of the traditional *RLC* configurations. It thus becomes a practical necessity to avoid the inclusion of coils in favor of *RC circuits that use ICs in active filters*. Such arrangements are both feasible and very effective, by making use of the flexible amplification provided by OP AMPs.

The objective of producing an effective filter circuit (one that is sharply tuned and does not have excessive losses) and still avoiding the use of coils can be accomplished in two basic ways:

1. By the use of a *frequency-selective RC network in the feedback and forward paths of an amplifier;* this approach is discussed next.
2. By the use of a *gyrator circuit*, which replaces the inductance *L* in the classical *RLC* circuits; this approach will be discussed in Section 12-6.

Passive Low-Pass *RC* Filter (Lag Network)

The passive low-pass *RC* filter of Fig. 12.4(a) consists of simply a resistor (R) in series with a capacitor (C) across which the output is taken. The cutoff frequency $(f_c = 16 \text{ kHz})$ at which the output of the filter is 70.7 percent of its dc value (or -3-dB down) is obtained from its transfer function as

$$f_c = \frac{1}{2\pi RC}, \quad \text{in hertz,}$$

where *R* is in ohms and *C* is in farads.

For a simple first-order filter as the one shown, with a single *RC* product, this -3-dB point occurs at the frequency where the reactance of the capacitor (X_c) is equal to resistance (R) and results in the widely used straight-line approximation of the Bode plot (with slope equal to -6-dB/octave).

The frequency depends (inversely) on the values chosen for *R* and *C*, and the resulting *RC* product, as shown in Table 5.1; thus the values of 2 kΩ and 0.005 μF (shown in the circuit of Fig. 12.4) give an *RC* product of $10 \times (10^{-6})$ or 10 μs, and a cutoff frequency f_c or $f_o = 16$ kHz.

When this passive low-pass type of *RC* filter is applied as frequency compensation for an OP AMP, it is known as a *lag network*. The shunt capacitor called for in the lag network is sometimes self-contained in the OP AMP, as is the case of the *internally compensated* 741 (and equivalent) types.

Modified Lag Network

In many other cases the lag network is made from external components. This is particularly the case where a modified lag network is used to shape the response curve, as in Fig. 12.4(b) and (c). By adding the resistor R_2 to the capacitor in the shunt position, the -6-dB/octave slope is modified so that the response

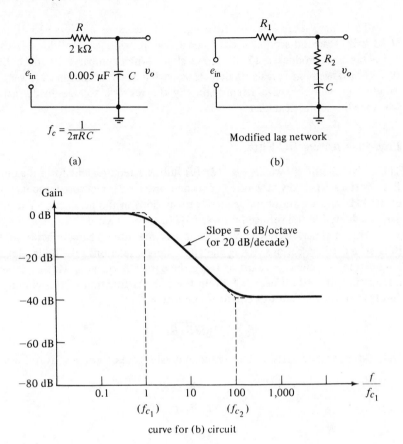

Figure 12.4 Passive low-pass filters (lag network): (a) simple circuit with R C values · for corner (or cutoff) frequency (f_c) of 16 kHz, with "roll-off" of 6-dB/octave; (b) modified circuit with two corner frequencies f_{c_1} and f_{c_2}; (c) normalized frequency response curve for modified lag network of (b), with $f_{c_2} = 100 f_{c_1}$.

tends to level off at the higher frequencies. With the addition of R_2, two corner frequencies are obtained,[2] as follows:

$$f_{c1} = \frac{1}{2\pi(R_1 + R_2)C},$$

$$f_{c2} = \frac{1}{2\pi R_2 C}.$$

[2] A. Barna, *Operational Amplifiers*, John Wiley & Sons, Inc. (Interscience Division), New York, 1971.

To achieve more effective filtering, say to reach as much as -15-dB down at 32 kHz, a second section needs to be added, providing an additional 6-dB for the octave, to produce a 12-dB/octave slope. When compared to an *LC* filter, this *RC* arrangement reveals its defects as having a lack of sharpness (a very low value of *Q*) and excessive attenuation (by the resistive voltage-divider action) when feeding a practical load.

Low-Pass Active *RC* Filter

The active circuit shown in Fig. 12.5(a) fulfills a requirement for a maximally flat (Butterworth) low-pass filter of second order with a cutoff frequency (f_c) of 10 kHz and a gain of unity (zero attenuation) in the passband.[3] The unity gain is achieved in the voltage-follower *LM110* type OP AMP.

This comparatively simple circuit has the filter characteristics of two isolated *RC* filter sections. Using the approximate relations of a Bode plot, the attenuation is roughly -12-dB at twice the cutoff frequency, with an ultimate attenuation of -40-dB/decade.[4] From the transfer function of this circuit, the cutoff frequency (f_c) of this circuit is obtained as[5]

$$f_c = 1/2\pi\sqrt{R_1 R_2 C_1 C_2},$$

and from this characteristic, the component values are determined as follows:

$$C_1 = \frac{R_1 + R_2}{\sqrt{2} R_1 R_2 (2\pi f_c)}$$

and

$$C_2 = \frac{\sqrt{2}}{(R_1 + R_2) 2\pi f_c}.$$

More complicated active filter circuits use more than a single OP AMP, along with multiloop feedback, to obtain desired characteristics, such as a higher *Q*. The *effect of the value of Q* on the low-pass function of a multiloop "universal" active filter (discussed in the next section) is shown in Fig. 12.5(b). This is also a second-order filter action (slope of -40-dB/decade), and also indicates that a *Q* value of about 0.7 achieves the maximally flat characteristics.

[3] National Semiconductor, LM110 data sheet.

[4] J. Eimbinder, ed., *Designing with Linear Integrated Circuits*, John Wiley & Sons, Inc., New York, 1969.

[5] Although the derivation of this (and more complicated) transfer functions is properly the task of texts in circuit analysis, a very straightforward derivation of this transfer function is given as an example in G. J. Deboo and C. N. Burrous, *Integrated Circuits and Semiconductor Devices*, 2nd edition, McGraw-Hill Book Company, New York, 1977.

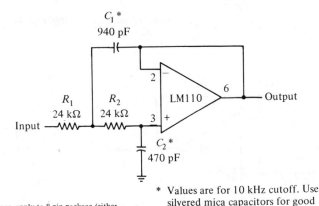

C_1 *
940 pF

R_1
24 kΩ

R_2
24 kΩ

2 −

3 +

LM110

6

Output

Input

C_2 *
470 pF

NOTE: Pin nos. apply to 8-pin package (either 8-pin metal can or 8-pin mini-DIP types).

* Values are for 10 kHz cutoff. Use silvered mica capacitors for good temperature stability.

(a)

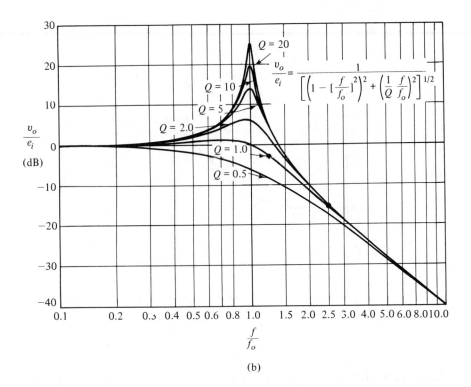

$$\frac{v_o}{e_i} = \frac{1}{\left[\left(1 - [\frac{f}{f_o}]^2\right)^2 + \left(\frac{1}{Q}\frac{f}{f_o}\right)^2\right]^{1/2}}$$

(b)

Figure 12.5 Low-pass active filter: (a) using voltage-follower LM 110 type of OP AMP (*National Semiconductor*), with values for high-frequency cutoff (f_c or f_o) of 10 kHz; (b) normalized frequency response plot for second-order low-pass active filter, showing effect of Q; value $Q = 0.7$ (interpolated) provides maximally flat response of Butterworth filter type. (*Beckman-Helipot Universal Active Filter, Model 881*)

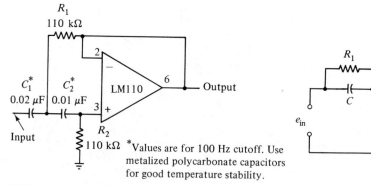

NOTE: Pin nos. apply to 8-pin package (either
8-pin metal can or 8-pin min-DIP types).

(a) − active   (b) − passive

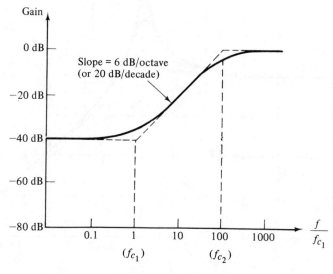

(c) − curve for (b) circuit

Figure 12.6   High-pass filters (lead network): (a) active high-pass filter, where
the position of resistors and capacitors of low-pass (Fig. 12.5) are
interchanged, and values are for low-frequency corner of 100 Hz
(*National Semiconductor*); (b) passive high-pass (lead) network;
(c) normalized frequency response curve for (b), with corner
frequency f_{c_2} equal to $100 f_{c_1}$.

Active High-Pass *RC* Filter (Lead Network)

A similar single-loop high-pass filter can be obtained by substituting capacitors for the resistors (and resistors for the capacitors) in the previous low-pass active filter. This high-pass circuit is shown in Fig. 12.6(a) with the values for a cutoff frequency (f_c) of 100 Hz. Here again, the circuit uses the *LM110* type of voltage-follower OP AMP.

The high-pass circuit is generally known as a *lead network*, which is shown in passive form in Fig. 12.6(b). This puts the filter capacitor effectively in series with the signal [as opposed to the shunt position of the low-pass (or lag) network], and causes a rising response with frequency. The lead network has a resistor placed across the capacitor in this case, which causes a leveling off at the higher frequencies as shown in Fig. 12.6(c). This modification causes *two corner frequencies* to appear, as follows:

$$f_{c1} = \frac{1}{2\pi R_1 C},$$

$$f_{c2} = \frac{1}{2\pi C[R_1 R_2/(R_1 + R_2)]}.$$

Band-Pass Active Filters

A band-pass filter is designed to pass a group (or band) of frequencies around some center frequency, while attenuating both the higher- and lower-frequency signals. One straightforward method for doing this is to *cascade a low-pass and high-pass filter.* This method has the advantage of attaining a very broad passband, and of separately tuning the high and low cutoff frequencies of the filter.

For use with a single OP AMP in a single-loop feedback, the circuit of Fig. 12.7 makes use of a twin-T network.[6] Analysis of the transfer function of the twin-T network shows that currents in the upper T and lower T are equal and of opposite phase. Accordingly, at the center frequency the twin-T network in the feedback loop passes no current, and so the center frequency is not attenuated. For frequencies on either side of the center frequency, however, feedback current progressively increases, causing the response to fall off on either side of the center frequency. Even with the use of the general-purpose 741 OP AMP, the passband can be made quite narrow. (This effect is the opposite of the *notch filters*, which use the twin-T network in the input, rather than in the feedback path, and hence can be made to have a very sharp rejection response. An example of a notch filter is shown in Fig. 12.8.)

[6] R. Melen and H. Garland, *Understanding IC Op Amps*, Howard W. Sams Company, Inc., Indianapolis, 1971.

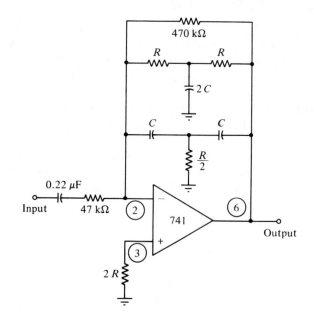

Figure 12.7 Band-pass active filter: a single-loop feedback circuit using a twin-T as the feedback element.

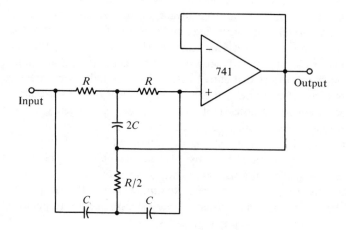

Figure 12.8 Basic notch (or *band-reject*) filter using a twin-T network at the input.

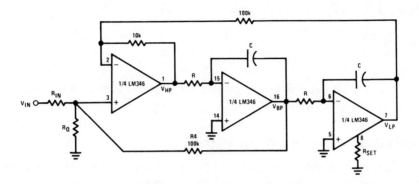

Circuit synthesis equations (for circuit analysis equations, consult with the AF100 and LM148 data sheet).

Need to know desired: f_o = center frequency measured at the BP output
  Q_o = quality factor measured at the BP output
  H_o = gain at the output of interest (BP or HP or LP or all of them)

▲ Relation between different gains: $H_{o(BP)} = 0.316 \times Q_o \times H_{o(LP)}$; $H_{o(LP)} = 10 \times H_{o(HP)}$

▲ $R \times C = \dfrac{5.033 \times 10^{-2}}{f_o}$ (sec)

▲ For BP output: $R_Q = \left(\dfrac{3.478\, Q_o - H_{o(BP)}}{10^5} - \dfrac{H_{o(BP)}}{10^5 \times 3.478 \times Q_o} \right)^{-1}$; $R_{IN} = \dfrac{\left(\dfrac{3.478\, Q_o}{H_{o(BP)}} - 1 \right)}{\dfrac{1}{RQ} + 10^{-5}}$

▲ For HP output: $R_Q = \dfrac{1.1 \times 10^5}{3.478\, Q_o\, (1.1 - H_{o(HP)}) - H_{o(HP)}}$; $R_{IN} = \dfrac{\dfrac{1.1}{H_{o(HP)}} - 1}{\dfrac{1}{RQ} + 10^{-5}}$

Note. All resistor values are given in ohms.

▲ For LP output: $R_Q = \dfrac{11 \times 10^5}{3.478\, Q_o\, (11 - H_{o(LP)}) - H_{o(LP)}}$; $R_{IN} = \dfrac{\dfrac{11}{H_{o(LP)}} - 1}{\dfrac{1}{RQ} + 10^{-5}}$

(a)

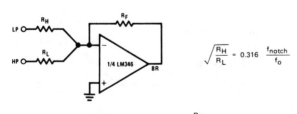

$\sqrt{\dfrac{R_H}{R_L}} = 0.316\, \dfrac{f_{notch}}{f_o}$

Determine R_F according to the desired gains: $H_{o(BR)} \Big|_{f \ll f_{notch}} = \dfrac{R_F}{R_L}\, H_{o(LP)}$, $H_{o(BR)} \Big|_{f \gg f_{notch}} = \dfrac{R_F}{R_H}\, H_{o(HP)}$

(b)

Figure 12.9 The LM346 quad OP AMP used in the realization of active filters: (a) basic noninverting state-variable active filter block; (b) for a notch (BR) output, the fourth amplifier is used to sum the LP and HP outputs of the basic filter. (*National Semiconductor*)

For either the pass or rejection function of the twin-T network, the design formulation for the center frequency f_o is the same:

$$f_o = \frac{1}{2\pi RC}, \quad \text{in hertz.}$$

The aid for determining component values in this formula by use of their *RC* product is given in Table 5-1 and also applies here.

Using Quad OP AMPs for Active Filters

The *National Semiconductor LM146/246/346* series of quad amplifiers consists of four independent, high-gain, internally compensated, low-power, programmable OP AMPs. By means of two external resistors, the user may program the amplifier for gain-bandwidth product, slew rate, supply current, and so on. The dc voltage gain is 120-dB, and the minimum gain-bandwidth product is 0.8 MHz. The power supply range is from ±1.5 to ±22 V, and overload protection is provided for the input and the output.

Figure 12.9(a) illustrates how three of the OP AMPs in the LM346 chip are used for a *state-variable active filter* that provides high-pass (HP), band-pass (BP), and low-pass (LP) filtering. Design equations also are included in the figure. A band-reject (BR), or notch, filter may be realized by summing the LP and HP outputs using the fourth OP AMP, as illustrated in Fig. 12.9(b). A simple BP, LP filter building block using three OP AMPs is given in Fig. 12.10, and Fig. 12.11 shows a three-amplifier notch filter.

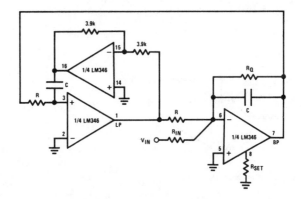

- If resistive biasing is used to set the LM346 performance, the Q_0 of this filter building block is nearly insensitive to the op amp's GBW product temperature drift; it has also better noise performance than the state variable filter.

Circuit Synthesis Equations

$$H_{o(BP)} = Q_0 H_{o(LP)}; R \times C = \frac{0.159}{f_o}; R_Q = Q_0 \times R; R_{IN} = \frac{R_Q}{H_{o(BP)}} = \frac{R}{H_{o(LP)}}$$

Figure 12.10 A basic BP, LP filter building block. (*National Semiconductor*)

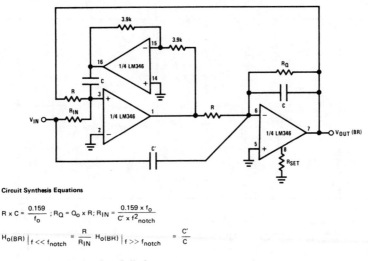

Circuit Synthesis Equations

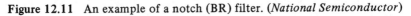

$$R \times C = \frac{0.159}{f_o} \quad ; R_Q = Q_o \times R; R_{IN} = \frac{0.159 \times f_o}{C' \times f^2_{notch}}$$

$$H_{0(BR)}\Big|_{f \ll f_{notch}} = \frac{R}{R_{IN}} \quad H_{0(BR)}\Big|_{f \gg f_{notch}} = \frac{C'}{C}$$

• For nothing but a notch output: $R_{IN} = R$, $C' = C$.

Figure 12.11 An example of a notch (BR) filter. (*National Semiconductor*)

12-6 GYRATORS AND *Q*-FACTOR IN ACTIVE FILTERS

In discussing the *RC* type of active filter, we should not neglect the massive storehouse of circuit-analysis theory that is based on the use of coils in the classical *RLC* filter. It is possible to *simulate the action of an inductor by a gyrator circuit*, which uses an active *RC* circuit to give the effects of inductor action. The gyrator effect can be shown most clearly in the circuit of Fig. 12.12, where two OP AMPs are used to provide a simulated inductor, whose simulated inductance is given by the formula[7]

$$\text{simulated inductance } L = \frac{R_1 R_2 R_4 C_1}{R_3},$$

where *L* is in henries, *R* is in ohms, and *C* is in farads.

As a simple illustration of this action,[8] if we assume that $R = R_1 = R_2 = R_3 = R_4 = 10 \text{ k}\Omega$ and $C_1 = 0.01 \text{ }\mu\text{F}$, we find

$$L = C_1 R^2 = 0.01 (10^{-6}) 100 (10^6)$$

$$= 1 \text{ H.}$$

[7]Melen and Garland, *Understanding IC Op Amps.*

[8]The derivation of this action, beyond the scope of this book, may be found in Deboo and Burrous, *Integrated Circuits and Semiconductor Devices.*

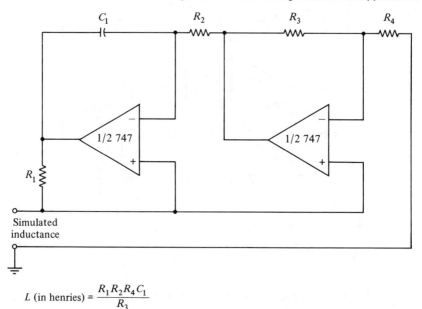

$$L \text{ (in henries)} = \frac{R_1 R_2 R_4 C_1}{R_3}$$

Figure 12.12 Gyrator circuit: with four equal resistors (R), the simulated inductance (L) equals $C_1 R^2$, in henries.

It should be noted that the inductance is simulated with one end of the input at ground, and therefore this kind of circuit cannot be used to simulate a "floating" inductance, where neither end is grounded. Another, more complicated method, is used, employing the concept of negative resistance to get around this difficulty, especially for low-pass active circuits, where the equivalent coil must be in a series position, and is therefore ungrounded.

Generalized Impedance Converter (GIC)

An example of a commercially available chip, called a *generalized impedance converter* (GIC), from which a gyrator may be realized is the *National Semiconductor AF120* GIC. In addition, a frequency-dependent negative resistance (FDNR) may be obtained. The AF120 contains a pair of OP AMPs and four premium thin-film resistors (Fig. 12.13). Its use as a gyrator for the realization of a floating inductance is illustrated in Fig. 12.14. The value of C_2 required for a given value of inductance L is

$$C_2 = \frac{L}{R_1 R_4}$$

where $R_1 = R_4 = 7,500 \ \Omega$. Hence $R_1 R_4 = 7,500^2 = 5.625 \times 10^7$. If, for example, an $L = 1$ H is required, $C_2 = 1/(5.625 \times 10^7) = 0.018 \ \mu F$.

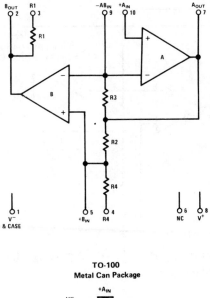

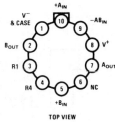

Figure 12.13 The AF120 generalized impedance converter (GIC): (a) simplified schematic; (b) connection diagram. (*National Semiconductor*)

Employing two capacitors, Fig. 12.15 shows how the AF120 is used in the realization of a frequency-dependent negative resistance, Z_i:

$$Z_i = \frac{-1}{\omega^2 R_1 C_1 C_5}$$

An example of the AF120 used to realize a third-order Butterworth LP filter is provided in Fig. 12.16; the cutoff frequency is 1,590 Hz.

Effect of Quality Factor (Q)

The sharpness of tuning in a filter circuit, especially for band-pass filters, is generally given by the relation for the -3-dB bandwidth [$BW(3\text{-dB})$ = width between the -3-dB points on the low and high sides of the center frequency

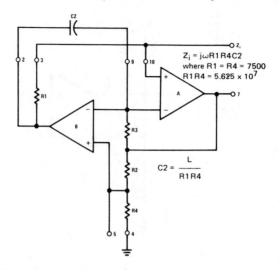

Figure 12.14 The AF120 used to gyrate a floating inductance. (*National Semiconductor*)

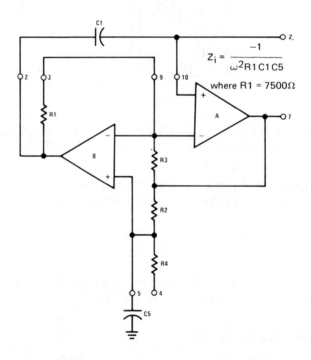

Figure 12.15 The AF120 used to realize a frequency-dependent negative resistance. (*National Semiconductor*)

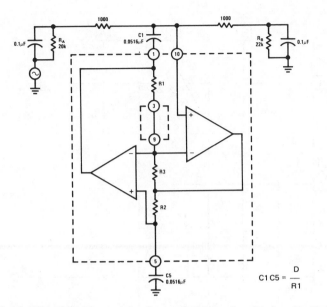

$$C1\ C5 = \frac{D}{R1}$$

Figure 12.16 The AF120 used in the realization of a third-order Butterworth low-pass filter. (*National Semiconductor*)

(f_o)] as follows:

$$BW\ (3\text{-dB}) = \frac{f_o}{Q}.$$

The effect of Q on the bandwidth of an active filter can be illustrated by an example of a universal active filter (*Beckman Helipot, model 881*). This hybrid circuit, Fig. 12.17(a) uses three OP AMPs (K_1, K_2, and K_3) in a multiloop feedback circuit, and K_4 as an uncommitted 741 amplifier for gains greater than unity. Functionally, K_1 is a summing amplifier, and K_2 and K_3 are integrators, providing simultaneous low-pass, high-pass, and band-pass outputs. Two external resistors set the desired cutoff frequencies (f_o), while three other external resistors determine gain (A_{VCL}) and the Q of the overall filter. The effect of the selected Q on the bandwidth of the band-pass filter is shown by the graph of Fig. 12.17(b), which is in normalized form for unity gain and center frequency (f_o). For sharp tuning, values of Q greater than 10 are generally used.

12-7 LOGARITHMIC AMPLIFIER

The "linear" OP AMP can be combined with a nonlinear element (such as the base-emitter diode of a transistor) to achieve the very useful nonlinear *loga-*

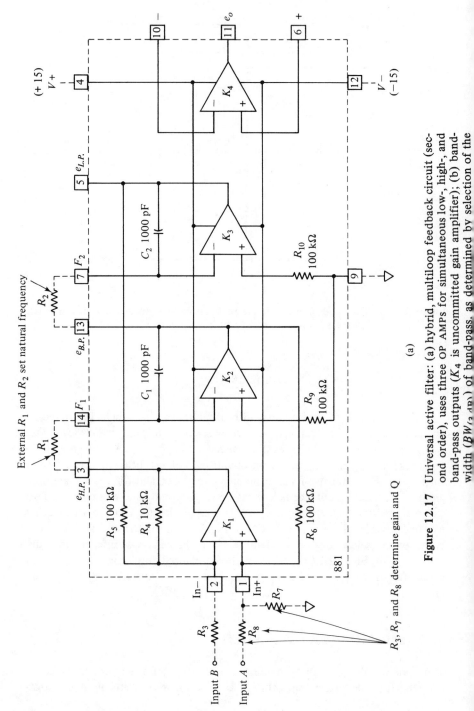

Figure 12.17 Universal active filter: (a) hybrid, multiloop feedback circuit (second order); uses three OP AMPs for simultaneous low-, high-, and band-pass outputs (K_4 is uncommitted gain amplifier); (b) bandwidth ($BW_{(3\,dB)}$) of band-pass, as determined by selection of the

Band-pass transfer characteristics normalized for unity gain and frequency

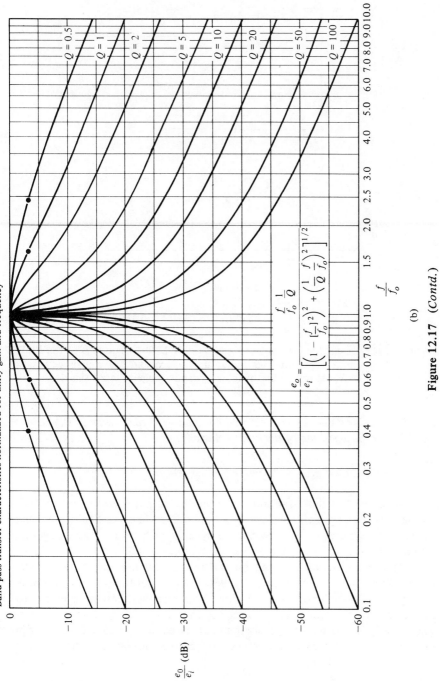

$$\frac{e_o}{e_i} = \frac{\frac{f}{f_o}\frac{1}{Q}}{\left[\left(1 - \left[\frac{f}{f_o}\right]^2\right)^2 + \left(\frac{1}{Q}\frac{f}{f_o}\right)^2\right]^{1/2}}$$

(b)

Figure 12.17 (*Contd.*)

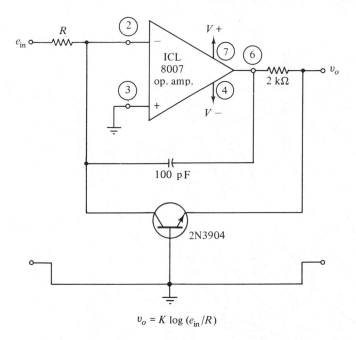

$$v_o = K \log (e_{in}/R)$$

NOTE: Pin nos. apply to 8-pin package (either
8-pin metal can or 8-pin mini-DIP types).

Figure 12.18   Logarithmic amplifier circuit: the ICL8007 OP AMP is of the FET-input type for high input impedance and has a high slew rate (6 V/μs) to handle the large dynamic range of input signals.

rithmic function (Fig. 12.18). In this circuit the *output voltage* (v_o) *is proportional to the logarithm of the input voltage* (e_{in}):

$$v_o \text{ varies as } \log (e_{in}).$$

As a result of this logarithmic relation, the output is greatly compressed, so the response of a meter across the output will act like a *decibel meter;* that is, the meter scale will cover a very wide dynamic range (such a scale might be −40 to +20-dB, a range of three orders of magnitude or 1,000:1).

The circuit action is based on the exponential relation of the base-to-emitter junction of the transistor (I_B versus V_{BE}) and the resultant feedback current that flows through the collector. Note that the transistor is employed in a common-base configuration, with the collector at a practically zero voltage, since it is tied to the virtual ground at the inverting terminal. Keeping in mind the fact that the OP AMP keeps the feedback current equal to the input current (e_{in}/R), we can express the output voltage $(v_o = V_{BE})$ as

$$v_o = K \, (\log e_{in}),$$

where the constant K includes the familiar exponential relation of diode leakage current and also the effect of fixed resistance R.

The OP AMP used (*Intersil ICL8007*) is of the FET-input type, which is internally compensated (pin compatible with the 741 type) and which, in addition, has the desirable properties of a very high input impedance (10^{12} Ω, or 1,000,000 MΩ) and an improved slew rate of 6 V/μs, thus allowing it to accommodate the wide range of input voltages that is necessary here.

For use as a decibel meter, the position of 0-dB on the scale can be arranged (by means of the offset-null provision) by applying an input voltage that corresponds to the 0-dB reference voltage (a much used reference level, the 0 dBm for 1 mW in 600 Ω calls for an input of positive 0.775 V).

The logarithmic circuit is also often adapted to use as a *speech compressor*. Employed in this way, it overcomes a basic limitation of microphone amplifiers that are required to respond to both low and high levels . In the presence of loud sounds, such amplifiers are prone to distort or, in the case of modulators, to overmodulate a transmitter. When the speech compressor is inserted between the microphone and its amplifier, the latter can be set to amplify a fairly low sound; then, when a loud sound is applied, the output level is compressed so that it is not that much greater, thus preventing overdriving of the amplifier.

Another interesting application, functioning as a special kind of *analog multiplier*, sums the output of two log amplifiers, which then feeds an antilog amplifier. (The circuit for the antilog amplifier is made very simply by trans-

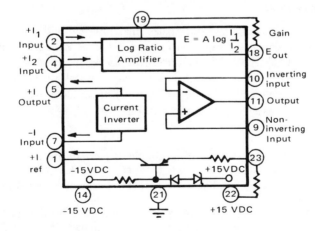

Figure 12.19 Block diagram of the 4127 log amplifier. (*Burr–Brown*)

posing the transistor into the input circuit and the resistor into the feedback circuit.) Then, when e_1 and e_2 are respective inputs to each log amplifier, the summer presents log e_1 + log e_2 to the antilog amplifier, which then produces the antilog of this sum, as the product of the two inputs[9]:

$$\text{antilog}\,(\log e_1 + \log e_2) = e_1 \times e_2.$$

An example of a packaged (double-wide DIP) hybrid log amplifier that accepts input signals of either polarity from current or voltage sources is the *Burr-Brown 4127* device. Developed for log conversions, it functions with good accuracy for up to six decades of input current and four decades of input voltage. A block diagram of the unit is provided in Fig. 12.19. The current inverter, coupled with a precise internal reference, allows preprogramming of the 4127 as a log, log ratio, or antilog amplifier. For greater versatility, an uncommitted OP AMP is contained in the package.

12-8 LOW-NOISE AND HIGH-IMPEDANCE (ELECTROMETER) APPLICATIONS

The problem of extracting a weak signal from various unwanted interfering effects (generally lumped together as noise signals) requires the selection of special types of linear ICs. These include *low-noise types* in those instances where the signal is very weak, or *electrometer types* in instances where a very high input impedance is necessary.

Low-Noise Types

Of the many types of unwanted signals that interfere with the amplification of the desired signal, the rejection of *common-mode signals* has already been dealt with in Chapter 11. While an intensive discussion of many other types of noise is beyond the scope of this book, the practical aspect of selecting an OP AMP that minimizes *burst (or "popcorn") noise*, along with the $1/f$ noise, can be presented as a straightforward method.[10]

---

[9] This and other advanced multiplying functions are described in G. E. Tobey, "Analog Modules," *Electronic Products*, Feb. 19, 1973.

[10] The following references are available in the commercial literature for more complete discussion of noise-reduction techniques:

    (a) For a general discussion of noise sources and shielding techniques, see Letzter and Webster, "Noise in Amplifiers," *IEEE Spectrum*, Aug. 1970.

For such noise reduction in the general-purpose *741* type, RCA offers premium type *CA6741*, which is tested for rigid low-noise standards for the low-burst-noise property and good performance for the $1/f$ noise, which is especially pronounced at the low frequencies.

In a similar manner, for the more refined programmable (or micropower) types of OP AMP that might be used for *low-level signals* (in medical electronics, for example), the *CA6078A* offers low-noise specifications that are improved over the general micropower *CA3078A* type.[11] Another example of a low-noise, OP AMP is the *Harris 909/911* type. Typical broadband noise (10 Hz to 1 kHz) is 1 μV.

Electrometer Types

Currents as low as a fraction of a picoampere (10^{-12} A) are typical of signals encountered in chemical (pH) and medical (EEG) electronics and, for such weak-signal measurements, require the special *very high impedance type of amplifier* known as an electrometer. This necessity arises from the fact that a very high input impedance is necessary in the device in order to develop a significant voltage drop from such a minute current, along with the fact that most of the sources being measured are themselves of the very high internal impedance type (as, for another example, the piezoelectric or capacitive type of transducer).

These conditions naturally suggest the use of field-effect transistors (FETs) to obtain the necessary high input impedance. The FET-input types of OP AMPs, having junction FETs (JFETs) integrated on the IC chip, satisfy this requirement, and examples of such monolithic forms will be presented, offering at least 100 GΩ (1 $\times$ 10^{11} Ω) and more for the input resistance and corresponding low leakage currents. [Where still better values are required, *dual* MOSFETs may be used externally; these can provide such values of input resistance as $R_{in} = 10^{15}$ Ω, and leakage currents as low as 1 fA (10^{-15} A)].[12]

(b) For signal-averaging and correlation techniques, see the *Hewlett-Packard Journal* of Apr. 1968 (on the 5480A signal analyzer) and of Nov. 1969 (on the 3721-A correlator).

(c) For phase-locking methods, see the Keithley 822 phase-sensitive detector and the *Princeton Applied Research Technical Bulletin 109* (for the JB-5 lock-in amplifier).

[11] *Op Amp Data Bulletins* (File Nos. 531 and 535) give additional application and test data (RCA).

[12] Discussed in greater detail for advanced electrometer use in Deboo and Burrous, *Integrated Circuits and Semiconductor Devices*.

Measuring Picoamperes

An electrometer circuit that can measure current in the picoampere (10^{-12} A) range is shown in Fig. 12.20. The OP AMP used is the *National* type *LH0052* having an input resistance (R_{in}) of 1,000,000 MΩ (1×10^{12} or 1 TΩ), paralleled by 4 pF; the input-offset current is in the range of femtoamperes (10^{-15} A). Since it is internally compensated, it is pin-compatible with the 741 type. The diagram shows the need for using shielded leads for the input probe and the very high feedback resistor of 100,000 MΩ (100 GΩ or 10^{11} Ω). This circuit is suggested for use in pH meters and radiation detectors, as examples of sensors that produce very small signal currents.

Guarding Circuits

In mounting electrometer-type OP AMPs, it is very possible for the leakage current across the sockets (or printed-circuit boards) to exceed the picoampere bias current of the device. Such leakage paths, which are essentially unbalanced, must be avoided because of the relatively large offset voltages that can be developed in the high impedances of the source and rest of the circuit. In using the *Burr-Brown model 3521* as an electrometer, for example, it is suggested that the inputs be connected to Teflon standoffs or, if mounted directly, to use a *guard pattern*, as shown in printed-circuit form in Fig. 12.21 for the metal-can package of the *3521*. By means of this guarding technique, when the guard pattern is connected as shown, leakage currents are rendered relatively harmless, since leakage paths from the board terminals are placed at practically the same potential as the input connection.

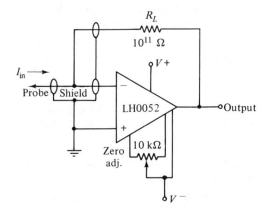

Figure 12.20 Picoampere (10^{-12} A) amplifier, used as an electrometer, suitable for pH and radiation detector applications. (*National Semiconductor*)

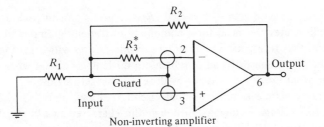

Non-inverting amplifier

* R_3 may be used to compensate
for very large source resistances
$(R_S > 5 \text{ M}\Omega)$, $R_3 = R_S$

Note: $\dfrac{R_1 R_2}{R_1 + R_2}$ must be LOW impedance

(a)

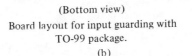

(Bottom view)
Board layout for input guarding with
TO-99 package.

(b)

Figure 12.21 Guard connections for very high input impedance amplifiers; minimizes effect of leakage currents at mounting terminals (in addition to conventional shielding of input and feedback leads, when amplifying weak signals). (*Burr–Brown, BB3521 series*)

Drift Considerations

In addition to the precautions mentioned as to shielding and guarding, it is important that the electrometer type have low-drift characteristics, even after the initial voltage offset has been nulled. These characteristics are carefully spelled out in the data sheets for the *LH0052* and *BB3521* examples cited. In the same category, the *Analog Devices AD523*, which has an initial bias current of less than 1 pA, is also carefully specified to account for any drift with temperature and/or with time.

12-9 SAMPLE-AND-HOLD APPLICATIONS

Among the digital-interface circuits mentioned in Chapter 10, the special type of sample-and-hold circuit serves as an example of a *storage application*, generally associated with analog-to-digital (A/D) conversion and peak-detector circuits. It illustrates the need for a combination of many desirable properties in one device, that is, especially good *input characteristics* with regard to high Z_{in}, low offset (V_{io}), and drift, all of which are difficult to obtain when *combined with high speed (or slew rate)* along with reasonable settling times.

A fairly simple sample-and-hold circuit is shown in Fig. 12.22(a), using the *LM 102* voltage follower. (Although the characteristics of $Z_{in} = 10^{10}$ Ω and a slew rate of 10 V/μs are good for a bipolar device, improvement in both figures can be obtained from FET-input types, such as the *AD528* with a slew rate of 50 V/μs.) In the circuit operation of part (a), the sampling signal closes

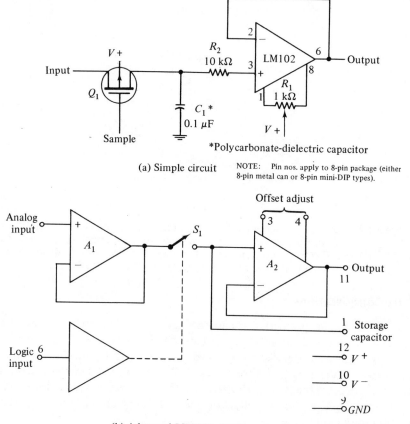

(a) Simple circuit   NOTE: Pin nos. apply to 8-pin package (either 8-pin metal can or 8-pin mini-DIP types).

(b) Advanced LH0043 circuit

Figure 12.22   Sample-and-hold circuits: (a) simple circuit with MOSFET sampling switch and LM102 voltage-follower; (b) advanced circuit of LH0043 OP AMP, self-contained in 12-lead TO-88 can, requiring only external storage capacitor. (*National Semiconductor*)

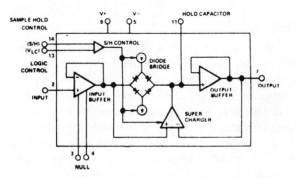

Figure 12.23 Block diagram of the SMP-11 sample-and-hold amplifier. (*Precision Monolithics, Inc.*)

the MOSFET switch (Q_1) and charges the storage capacitor C_1 to the input value. When the switch is opened, the voltage-follower action makes and holds the output equal to the sampled input. As a necessary consequence, the dielectric of the holding capacitor must be of a high-quality type, such as Teflon, polycarbonate, or the like.

For the higher-performance type of device, a complete sample-and-hold circuit that includes all the required buffering and FET sampling-gate elements in a single 12-lead, TO-8 metal-can package is contained in the *LH0023* OP AMP (or, for inverted logic, in the *LH0043* OP AMP). The functional circuit of the elements enclosed in the *LH0043* package (except for storage capacitor) is shown in part (b) of the figure. This represents a significant size reduction from the more conventional (and costly) discrete or module designs.

Another example is the *Precision Monolithics, SMP-11* sample-and-hold amplifier (Fig. 12.23). Employing two buffer amplifiers (input resistance of each is greater than 3×10^{10} Ω), the *SMP-11* is a noninverting unity gain device. A diode bridge is used for the switch and up to 50 mA charging current is available to the hold capacitor. The adjustable logic input threshold makes the device compatible with various logic families.

12-10 MULTIPLE (TRIPLE AND QUAD) OPERATIONAL AMPLIFIERS

The integration of three (and four) operational amplifiers on one chip affords great flexibility in a variety of special applications. Its use in the realization of active filters was considered in Section 12-6. In this section, the subject of multiple OP AMPs is considered in greater detail.

Triple OP AMP

The *CA3060*, for example, consists of three operational amplifiers of the operational-transconductance-amplifier (OTA) type, previously described in discussing the single *CA3078*/A of that type. In such OP AMPs, the transfer function is best described in terms of the ratio of its current output against the input voltage, that is, in terms of its *transconductance* (g_m); this arises from the fact that, unconventionally, the output impedance of this amplifier type is of a high value, and also because its resulting transconductance can be varied over a wide limit by the choice of an external resistor that changes the amplifier-bias-current (I_{ABC}) value.

Thus, by varying the external resistor (R_{ABC}), the bias current can be made to vary from 1 to 100 μA, with *corresponding g_m values of 300 to 30,000 μS.* This biasing is independent for each of the three amplifiers, and the functional schematic diagram in Fig. 12.24 shows the circuit of one of them. (Also shown in the diagram is the self-contained zener-diode system for regulating the bias-

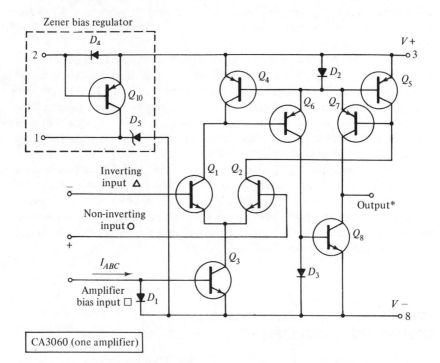

Figure 12.24 Circuit of one of the three amplifiers in the triple operational transconductance amplifier (OTA), CA3060; the amplifier bias current (I_{ABC}) of each independent amplifier in the 16-pin DIP package can be varied by a separate external resistor. (*RCA Solid State Division*)

current supply.) All the circuitry for the three amplifiers and the zener bias regulator is contained in a 16-lead DIP package.

Aside from the variable-bias system and the high output impedance, each of the three amplifiers has characteristics similar to the conventional OP AMP when feeding a relatively high resistance load. Some of the versatile applications of the *CA3060* include multiple-amplifier circuits for active filters, micropower, and various mixing or modulation functions.

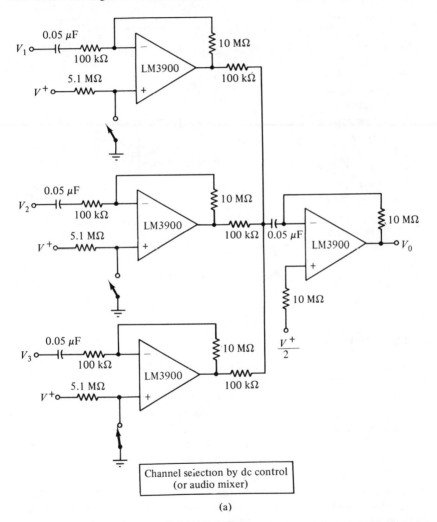

(a)

Figure 12.25 Quad amp (LM3900) applications: (a) the four independent OP AMPs of the 14-lead DIP package used as a three-channel mixer (or channel selector); (b) samples of waveforms obtainable from each quarter of the LM3900. (*National Semiconductor*)

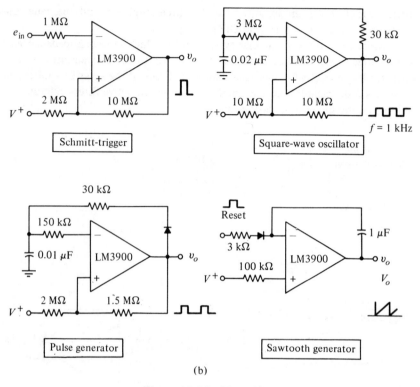

(b)

Figure 12.25 (*Contd.*)

Quad OP AMPS

The inclusion of four OP AMPs in one IC device is featured by the *LM3900*, the *MC3301/3401*, and other quad OP AMPs.

The *LM3900*, for example, consists of four independent internally compensated amplifiers, each similar to a typical dual-input OP AMP. It is packaged in a 14-lead DIP (8 input, 4 output, and 2 power-supply leads), and is designated to operate from a single power supply of 4 to 36 Vdc. Among the many possible configurations, Figs. 12.25(a) and (b) illustrate cases utilizing all four amplifiers of the device. Part (a) shows the circuit of an audio mixer (or channel selector) for three separate amplifier channels feeding the fourth amplifier as a mixer to furnish the desired output. In part (b) each quarter of the IC is used to generate a different output signal, as shown, producing, respectively, a Schmitt trigger, square-wave oscillator, pulse, and sawtooth generator, all from the single device.

Other applications are suggested by the circuit diagrams of *over 40 typical*

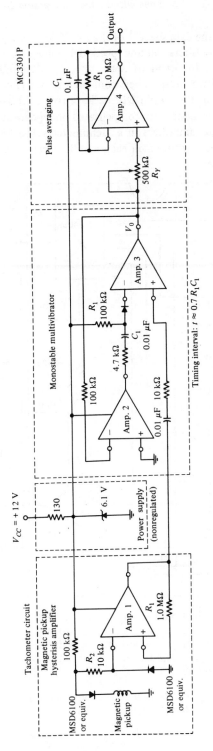

Figure 12.26 Quad OP AMP (MC3301/P), used in an automatic tachometer system; each of the four internally compensated independent amplifiers is employed for its particular function. (*Motorola Semiconductor Products, Inc.*)

applications included in the data sheet[13] of the *LM3900*. Among other suggested applications are *RC* active filters and low-speed high-voltage digital logic gates.

Another version of the quad OP-AMP type, the *MC3301P*, is specifically designed for single positive supply operation, such as in +12-V automotive applications. It is offered in the 14-pin DIP package (or as *MC3401P* in a 16-pin DIP package) with four internally compensated independent amplifiers. An application using all four amplifiers in a tachometer system is shown in Fig. 12.26, illustrating the ease with which each amplifier can be used functionally as part of an overall system.

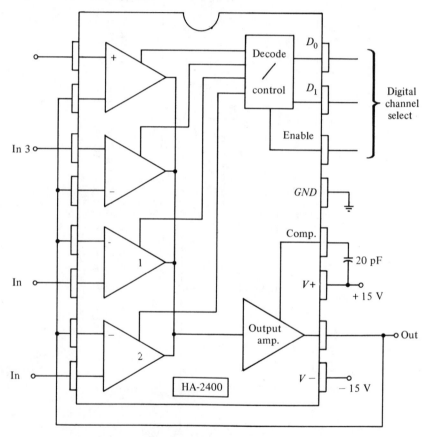

Figure 12.27　Gate-controlled monolithic quad OP AMP (HA2400): in this buffered multiplexing application, the four amplifiers, used as buffers, are combined with analog switching by digital selection and an output amplifier, all in one monolithic package. (*Harris Semiconductor*)

---

[13] *Linear Integrated Circuits*, a manual commercially available from National Semiconductor (see address in Appendix IV).

Another example[14] illustrating the integration of multiple OP AMPs in a single monolithic 16-lead DIP is furnished by the *Harris HA2400*, a gate-controlled quad OP AMP. In the buffered multiplexing application shown in Fig. 12.27, the feedback signal places the selected amplifier channel in a voltage-follower configuration. The decode-control terminals are used for channel selection. As a result of this interfacing function, the single quad OP AMP package takes the place of four input buffer amplifiers, four analog switches with digital decoding, and one output buffer amplifer.

12-11 VOLTAGE-CONTROLLED OSCILLATOR (VCO) AND WAVEFORM GENERATORS

While the main function of most linear ICs is generally regarded as some form of amplifier action, it is well to keep in mind that OP AMPs (and other LIC types) are equally useful as *oscillators*.

RC Oscillators

As was illustrated in the quad OP AMP of Fig. 12.25, any single OP AMP can easily be configured to generate *square waves, sawtooth waves,* or *pulses* with a minimum amount of external components. Additionally the production of *sine waves* is also accomplished easily in the popular *Wien-bridge circuit* shown in Fig. 12.28(a), which uses the internally compensated *LM107* (or its equivalent 741 type of OP AMP). In part (b) two OP AMPs are used to provide a dual-wave function generator, which produces a *triangle-wave* output by means of using the second OP AMP (*LM107*) to integrate the square-wave output from the first OP AMP (*LM101A*).

Voltage-Controlled Oscillator

All the preceding generators, it will be noted, employ *RC circuits*, where the frequency is determined by the *RC* time constant. [This can be seen particularly in the function generator of Fig. 12.28(b), where resistor R_3 is labeled as the frequency control.] However, there are many applications where the need is to control the frequency by means of an applied (or control) voltage. This function is achieved in the *voltage-controlled oscillator* (VCO). A typical VCO (*SE/NE566*) is illustrated in block-diagram form in Fig. 12.29(a), and in a connection diagram in part (b).

In this circuit, the external capacitor (C_1) is alternately charged in a linear fashion by one current source and then linearly discharged by another to pro-

---

[14] "Analog Switching without FETs," *Electronic Products*, Oct. 18, 1971.

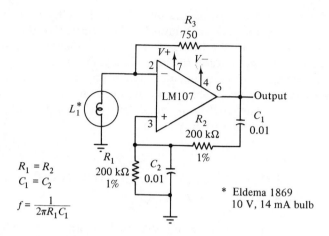

$R_1 = R_2$
$C_1 = C_2$

$f = \dfrac{1}{2\pi R_1 C_1}$

* Eldema 1869
10 V, 14 mA bulb

NOTE: Pin nos. apply to 8-pin package (either 8-pin metal can or 8-pin min-DIP types).

(a) Wien-bridge sine-wave oscillator

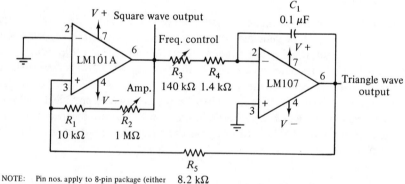

NOTE: Pin nos. apply to 8-pin package (either 8-pin metal can or 8-pin mini-DIP types).

(b) Function generator

Figure 12.28 Additional RC oscillators [see also Fig. 12.26 (b)]: (a) sine-wave oscillator, using the frequency selective $R_1 C_1$ elements of the Wien-bridge circuit; (b) two OP AMPs provide square- and triangular-wave outputs. (*National Semiconductor, Linear Applications, AN-31*)

vide the buffered triangular output. The Schmitt-trigger action determines the charge-discharge levels and provides the buffered square-wave output.

The connection circuit for the VCO in Fig. 12.29(b) shows the $R_1 C_1$ time constant that determines the free-running frequency. Additionally, the control voltage (V_C) at terminal 5 is set by the voltage divider consisting of

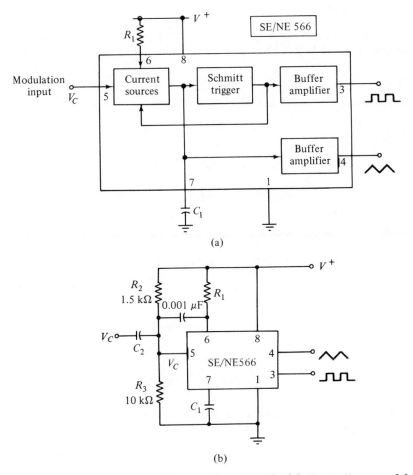

Figure 12.29 Voltage-controlled oscillator (VCO): (a) block diagram of function generator SE/NE566, whose frequency can be controlled by control voltage V_C; (b) connection diagram for initial bias at V_C (terminal ⑤) and $R_1 C_1$ elements for simultaneous production of square and triangular waves. (*Signetics*)

R_2 and R_3. The initial bias for the control voltage (V_C) must be set in the range between $\frac{3}{4}$ V+ and V+; for example, for a 12-V supply, V_C may vary between 9 and 12 V and thus exerts its influence in determining the output frequency. In all, then, there are three frequency-determining elements (R_1, C_1, and V_C); when keeping C_1 constant and at a fixed V_C, the frequency is *adjustable over a 10:1 frequency range*, by choice of R_1, up to a maximum frequency of 1 MHz. In a similar manner the frequency can be modulated over a 10:1 range by the control voltage V_C.

VCO Design Values

As an example, we may use a typical single-supply voltage of +12 V; using the values given in Fig. 12.29, we have

$$V_C = V \frac{R_3}{R_2 + R_3}$$

$$= 12 \frac{10 \text{ k}\Omega}{11.5 \text{ k}\Omega} = 12(0.87)$$

$$= 10.4 \text{ V}$$

This allows a V_C variation of -1.4 V (down to +9 V), and an upward variation of $+1.6$ V (up to +12 V).

Using round-number approximations, we shall choose such a control voltage (V_C) centered at 10.5 V (to allow a variation of ± 1.5 V at V_C). The approximate formula for the frequency (f_o) is

$$f_o \approx \frac{2(V^+ - V_C)}{V^+} \cdot \frac{1}{R_1 C_1},$$

and for this case

$$f_o \approx \frac{2(12 - 10.5)}{12(R_1 C_1)} = \frac{1}{4 R_1 C_1}$$

Since the recommended value for R_1 is in the range between 2 and 20 kΩ, we shall use $R_1 = 4$ kΩ. Then, for a nominal frequency (f_o) of 10 kHz, with $R_1 = 4$ kΩ,

$$C_1 = \frac{1}{16(10^3)\,10(10^3)} \quad \text{or} \quad \frac{1}{160(10^6)}$$

$$= 0.0062 \text{ }\mu\text{F} \quad \text{or} \quad 6,200 \text{ pF}$$

With these design values, V_C can be now varied ± 1.5 V to change the nominal frequency of 10 kHz; with V_C increased by $+1.5$ V (to 12 V), the output frequency decreases to about 2 kHz; and with V_C decreased by -1.5 V (to 9 V), the output frequency doubles to 20 kHz. Thus, as a result of this 3-V swing in V_C, the output frequency varies (or sweeps) over a 10:1 range between 2 and 20 kHz around a center frequency (f_o) of 10 kHz.

In a similar manner, the center frequency can be shifted from 10 kHz by changing R_1 from its initial value of 4 kΩ down to 2 kΩ (for a new higher $f_o = 20$ kHz), or we can increase R_1 by 5 times to 20 kΩ (for a new lower $f_o = 2$ kHz).

Consequently, for the values given (and within the 1 MHz maximum for this VCO), the available frequencies can be shifted by these particular choices from one fifth of 2 kHz (or 400 Hz) up to 40 kHz (in either a square or triangular waveform).

The ability to vary output frequency by means of an input voltage naturally suggests the use of the VCO for a *sweep frequency*, as might be used for testing (and displaying) the frequency response of an amplifier under test. Similarly, the VCO output provides a convenient method for *frequency-modulation* (FM) applications. Still another prominent development is the use of the VCO as an important element in the widely used *phase-locked-loop* (PPL) circuits (which were discussed in Section 8-8).

Waveform Generators

The integration of a *complete signal-generator capability* has been successfully accomplished in a single DIP package, as exemplified in the 14-lead *Intersil ICL8038*, and in the *16-lead XR205 model from Exar Integrated Systems*.

The functional block diagram of the *ICL8038* [Fig. 12.30(a)] shows the internal elements, including two current sources, with two associated comparators and a flip-flop. The current sources (I and $2I$) provide equal charge and discharge times to external capacitor C, producing the *triangular wave;* simultaneously, the flip-flop produces the *square wave.* Each of these waveforms has its own buffer amplifier to provide isolated outputs. The sine-converter section (itself composed of 16 transistors) functions in a nonlinear manner to change the triangular input into sine waves by providing a decreasing shunt impedance as the potential of the triangle moves toward the two extremes.

In addition to the three basic waveforms shown, the production of a wide-range (1,000:1) *sweep output* is shown by the connections in Fig. 12.30(b). The frequency (or repetition rate) for the various waveforms is selected with a minimum of external components. The stable waveforms cover a range of operation from below 0.001 Hz up to a bit more than 1 MHz.

Another type of monolithic waveform generator (*model XR205*) is shown in system block form in Fig. 12.31(a). The internal elements in this model are seen to include an analog multiplier that connects externally to an uncommitted VCO and a separate buffer amplifier. While suitable for split-supply operation, the external connections for single-supply operation are shown in part (b).

The outputs include the three basic *sine, square,* and *triangular* waveforms, and also provision for *ramp* or *sweep voltages*. In addition, a variety of *modulation outputs* can be obtained, including AM, FM, phase-shift-keyed (PSK), frequency-shift-keyed (FSK), and tone-burst outputs. (The XR205 data sheet shows a comprehensive series of oscilloscope photos illustrating the large variety of modulated outputs available.)

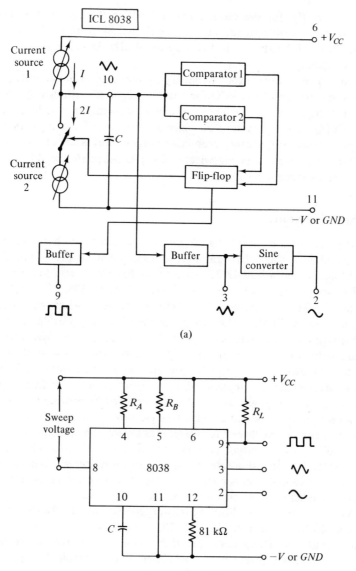

Figure 12.30 Waveform generator (ICL8038) in 14-pin DIP package: (a) functional block diagram, showing internal elements; (b) external connections for producing basic square, triangular, and sine waves. (*Intersil, ICL8038, Application Bulletin, AO12*)

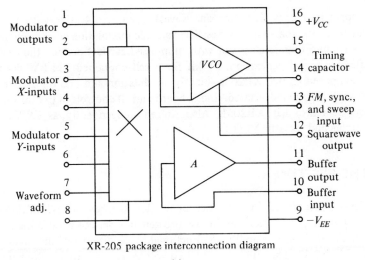

XR-205 package interconnection diagram

(a)

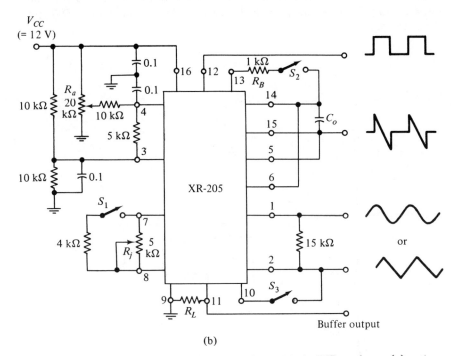

(b)

Figure 12.31 Waveform generator (XR205), in 16-pin DIP package: (a) system block diagram, showing analog multiplier, VCO, and buffer amplifier elements; (b) external connections for single-supply operation and basic waveforms (not including modulated wave outputs). (*Exar Integrated Systems, XR205 Data Bulletin*)

The upper frequency limit (for sine waves) extends to 4 MHz, with 20-ns rise and fall times for the square waves. The versatile modulation characteristics (which also include double sideband with suppressed carrier) provide the basis for very flexible signal generation. (A kit for a self-contained AM/FM signal generator, composed of two XR205s—one for modulation and one for carrier— is provided with a prepared printed-circuit board and all controls in the form of a Waveform Generator Kit, XR205K. Also offered independently, as *XR2207*, is a highly stable and versatile VCO.)

12-12 THE 555 TIMER

The *555 timer*, introduced by Signetics in the 1970s (*Signetics NE/SE555*), is a versatile building block that can be used to generate various waveforms and long time delays with good stability. A functional block diagram of the timer is provided in Fig. 12.32. The chip contains two voltage comparators, a discharge transistor, a resistive divider composed of three equal (R) resistors, a flip flop, and a totem-pole output stage. A dual version of the 555 timer, the *NE/SE556*, also is available.

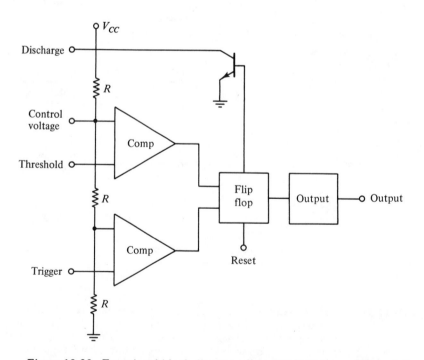

Figure 12.32 Functional block diagram of the 555/556 timer. (*Signetics*)

The comparator levels are set by the three equal resistors, which form a voltage divider with supply voltage V_{CC}. Hence, the threshold comparator is set at a reference voltage of $V_{CC}(2R)/(3R) = 2V_{CC}/3$ V, and the trigger comparator is set at $V_{CC}(R)/(3R) = V_{CC}/3$ V. Each comparator is connected to the flip flop. When the trigger input voltage is reduced to less than $V_{CC}/3$ V, the comparator changes its state and the flip flop switches to its high (1) state. External to the threshold terminal is an RC timing network. When the voltage across the threshold terminal exceeds $2V_{CC}/3$ V, the threshold comparator resets the flip flop to its low (0) state. For this condition, the discharge transistor is acting like a closed switch and discharges the timing capacitor connected to the threshold terminal. The timing cycle is now completed.

Applications for the 555 timer are legion; a few of the basic applications are examined in this section. A monostable (one-shot) multivibrator using the 555 timer is shown in Fig. 12.33. Requiring two external components (R_A and C), the trigger is normally a short negative pulse. The length (or width) of the

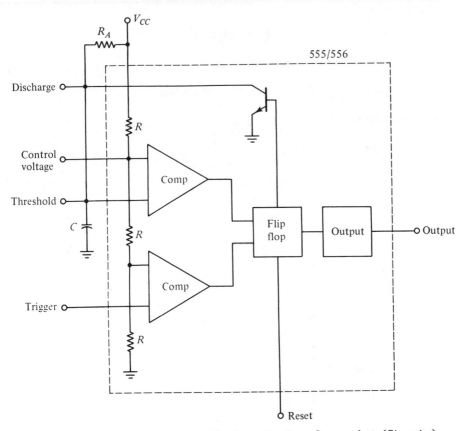

Figure 12.33 The 555 timer used in the realization of a one-shot. (*Signetics*)

output pulse, T, is

$$T = 1.1 R_A C \text{ s}$$

where R is in ohms and C in farads. If, for example, $R_A = 1,000\ \Omega$ and $C = 0.01$ μF,

$$T = 1.1 \times 10^3 \times 0.01 \times 10^{-6} = 0.011 \times 10^{-3} \text{ s} = 11 \text{ ms}$$

The addition of another resistor, R_B in Fig. 12.34, results in an astable (free-running) multivibrator that generates rectangular waveforms. During the interval that capacitor C charges from $V_{CC}/3$ to $2V_{CC}/3$ V, the output of the flip flop is in its 1 state. This interval, t_1, is equal to

$$t_1 = 0.695(R_A + R_B)C.$$

When the flip flop is in its 0 state, capacitor C discharges from $2V_{CC}/3$ to $V_{CC}/3$ V in the interval t_o:

$$t_o = 0.695 R_B C.$$

The period, T, of the rectangular waveform is

$$T = t_1 + t_o = 0.695(R_A + 2R_B)C$$

and the frequency of oscillation, f, is

$$f = \frac{1}{T} = \frac{1.44}{(R_A + 2R_B)C} \text{ Hz.}$$

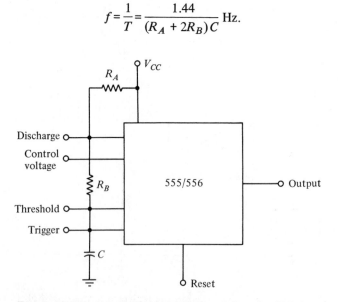

Figure 12.34   An astable multivibrator using the 555 timer.

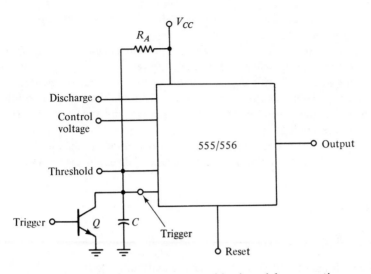

Figure 12.35 The 555 timer used in time-delay operation.

For example, if $R_A = R_B = 10$ kΩ and $C = 0.1$ μF,

$$f = \frac{1.44}{30 \times 10^3 \times 10^{-7}} = 480 \text{ Hz.}$$

The 555 circuit used to provide a time delay is illustrated in Fig. 12.35. In the time delay mode, the output does not change state upon triggering, as in the one-shot, but at some specified time after the trigger is received. To achieve this, the threshold and trigger terminals are connected and monitor the voltage across capacitor C, which is connected across transistor Q; the discharge function is not used. When Q is turned on, capacitor C is shorted to ground; the trigger signal therefore sees a low state and the output is high. When the capacitor reaches the threshold level, the output goes to its low state and remains there until Q is again turned on. The delay time, T, is given by the same expression as for the one-shot: $T = 1.1 R_A C$ s.

12-13 MULTIPLIERS, DIVIDERS, SQUARERS, SQUARE ROOTERS

The multiplying function (along with its modifications) originally developed for analog computation has become an important linear IC application, especially since the introduction of an entire analog multiplier contained on a single monolithic chip. The use of a four-quadrant multiplier (*MC1596*) was previously described (Section 8-7), where it served as a modulator (or mixer) in communication circuits. Its opposite use as a demodulator (or product detector) was also introduced in that section.

In this section, additional specialized applications will be described, using the *AD530* as a typical analog multiplier of the versatile type (Fig. 12.36). The analog multiplication in this device is based on the principle of *variable trans-conductance*, and serves many useful purposes. In addition to the multiplying function, it has the capability of dividing, squaring, and square rooting (hence its MDSSR designation); moreover, it also finds use under other functional names as previously mentioned in connection with modulation and demodulation, all of which makes it almost as versatile an LIC device as the OP AMP.

As a Multiplier

The multiplier circuitry depends essentially on its action in changing the gain of two sets of differential amplifier pairs by varying the emitter current of each pair. The multiplier transfer function yields an output (v_o):

$$v_o = \frac{XY}{10}.$$

The connection diagram for the multiplier [Fig. 12.36(a)] shows an adjustable scale control (5 kΩ) for the Y input, while the X input is applied directly. Provisions for zeroing the X and Y input, and also the output, are provided (when needed) by 20-kΩ pots. These offsets, which affect the dc accuracy, are dependent on close matching within each differential pair. By the use of modern IC technology and careful processing, the overall dc accuracy is kept down to production grades of ±2 and ±1 percent. The amplitude response with frequency (maximum undistorted output of ±10 V) is down 3-dB at 1 MHz.

It is interesting to compare this method of multiplying X and Y with the older *quarter-square method*, which can be implemented with a number of amplifiers acting as summing, difference, and squaring amplifiers, with the squarer using the principle of square-law response. This method was widely used in the past, based on the equation

$$(X + Y)^2 - (X - Y)^2 = X^2 + 2XY + Y^2 - X^2 + 2XY - Y^2 = 4XY.$$

The remaining $4XY$ is then divided by 4 to yield XY (accounting for its name of the quarter-square method). This method, although accomplishing its purpose, seems tortuously involved compared to the more direct method using the LIC multiplier to yield XY/K, which provides much simpler external circuitry for not only multiplying, but also for dividing, squaring, and extracting the square root, as shown next.

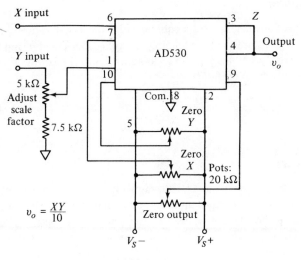

(a) Multiplier

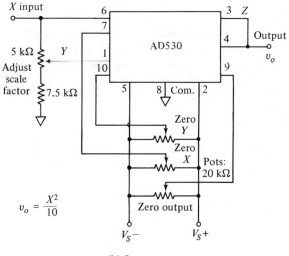

(b) Squarer

Figure 12.36 Analog multiplier (AD530): (a) connected as a multiplier to yield a product $XY/10$; (b) with X and Y inputs tied together, it is connected as a squarer, yielding $X^2/10$ as an output. (*Analog Devices, AD530 Data Bulletin*)

As a Squarer

The connection diagram for a squarer [Fig. 12.36(b)] shows that the squaring function is accomplished very simply by tying the X and Y inputs together. It is done here by connecting the scale-adjust control (for the Y input) to the X input, and yields an output (v_o):

$$v_o = \frac{X^2}{10}.$$

As a Divider

In the divider mode of the *AD530*, shown in Fig. 12.37(a), the numerator input signal is fed to the Z input and the denominator to X. The output terminal is connected to the Y input by way of the scale-adjust control. The transfer function for normal operation (where positive X signal is first inverted) yields an output (v_o):

$$v_o = \frac{10Z}{X}.$$

It is necessary to invert a positive X signal, since positive X as the denominator in this mode would cause the feedback to become positive and cause an overrange (or "pegging") output. Although not harmful to the device, the normal division operation is done with an inverted positive X signal.

As a Square-Rooter

The connections for the square-root function are shown in Fig. 12.37(b). Here the X and Y inputs are tied together, as in the squaring function, and the connections are arranged so that the Z input is divided by the squared magnitude in the output. Normal operation in this mode requires a positive input for Z, and produces a negative value for the output (v_o):

$$-v_o = \sqrt{10Z}.$$

Analog Multiplier Models

In module form, the four-quadrant multiplier may be obtained as a module that requires an external OP AMP for dividing (*Teledyne/Philbrick model 4450*) or one in which the OP AMP for this purpose is self-contained (*T/P model 4452*).

The use of the operational-transconductance-amplifier (OTA) type, such as the *CA3080/A*, has been previously mentioned in Section 11-8. It can also be

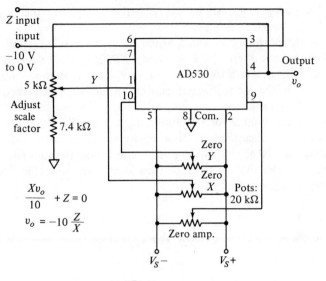

(a) Divider

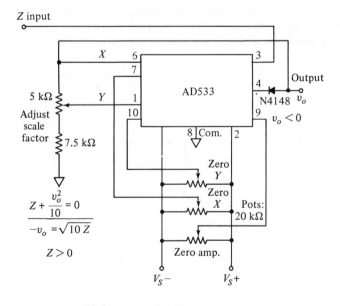

(b) Square-root function

Figure 12.37 Additional analog multiplier functions: (a) connected as a divider, to yield $10Z/X$; (b) connected for a square-root function to yield $\sqrt{10Z}$. (*Analog Devices, AD530 Data Bulletin*)

used as an analog multiplier by applying the Y input to the amplifier-bias-control (I_{ABC}) terminal.[15]

A model providing increased bandwidth as a versatile multiplier in a 16-lead DIP package is offered by *Exar, model XR2208*. Since the device contains a buffer amplifier plus an uncommitted OP AMP, in addition to the four-quadrant multiplier, all in the one package, it reduces the number of external components required in many instances.

The general symbol for a multiplier is shown in Fig. 12.38(a) and its use in a *mean-square* circuit in (b). Mean-square values are often useful for evaluating nonsinusoidal signals and can be obtained using the multiplier in a squaring mode, followed by an OP AMP used as an integrator [Fig. 12.38(b)].[16] The output v_o is given as

$$v_O = \cfrac{1}{KR_1C \displaystyle\int_0^t (e_{\text{in}})^2 \, dt}$$

Root-mean-square (true rms) measurement can be obtained by applying this mean-square value to the square-rooter connection shown previously.

The special applications shown here do not begin to exhaust the analog multiplier possibilities, which also include function generation, automatic gain

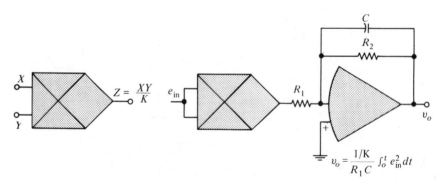

(a) Multiplier symbol   (b) Mean-square function

Figure 12.38   Symbol representation of analog multiplier: (a) general symbol for multiplier; (b) circuit used for a mean-square function, yielding an average value of e_{in}^2.

[15] Details are given in IC Application Note (ICAN-6668) in the *RCA Solid State Databook Series: SSD-202A*.

[16] J. Pepper, "Analog Multiplier Applications," *Instruments and Control*, June 1972.

control (AGC), and phase-comparator functions, among others. This versatility accounts for the presence of so many multiplier models, as is also the case for the OP AMP segment of the linear ICs.

For clearing up some of the possible confusion resulting from the multiplicity of models (as, in this instance, *MC1596, AD530, 4452, CA3080/A*, and *XR2208*), attention is called to the cross-referenced index in Appendix III, which identifies the manufacturers' designations and describes the LIC type. Appendix IV gives the addresses for obtaining the valuable technical data offered by the manufacturers for the specific models.

12-14 DIGITAL-VOLTMETER SYSTEM

As an example of a system application, the popular digital voltmeter (DVM) serves as a good illustration, showing how linear ICs are combined with digital-interface and digital ICs, all forming an operating voltmeter system.

The heart of the most widely used type of DVM is an interface linear IC, an A/D converter that uses a *dual-slope integration method* considered previously in Section 10-6. This method operates essentially as a voltage-to-time conversion, but it accomplishes the conversion in a manner that is significantly better than a simple integrator. As opposed to the linear ramp output that is obtained from a single integrator, the dual-slope method produces a time measurement that encompasses both the charge and discharge times of an integrating capacitor, and consequently it produces an output that is more accurately proportional to the ramp slopes.

The operation of the dual-slope method of integration is reviewed in Fig. 12.39. At the start of each measuring cycle, the unknown input voltage (e_{in}) is applied to a buffer amplifier, before being integrated by the OP-AMP integrator. With C_1 in the feedback loop, the capacitor is charged by e_{in}, producing a rising ramp for a fixed time (T_1), which is set for a predetermined amount of clock pulses. When this fixed number of pulses has been counted, the control circuits open switch S_1 and close either the S_2 or S_3 switch to apply a reference voltage ($+V_r$ or $-V_r$) of the opposite polarity to the input e_{in}. This discharges capacitor C_1 to produce the descending ramp, during which time the number of pulses is counted as T_2 until the integrator output falls to zero, and the control circuit (zero detector) stops the count. Since the charge time (T_1) is fixed, the discharge time (T_2) caused by the known reference voltage (V_r) is determined solely by the charge acquired from the unknown voltage (e_{in}). For example, if the charge time is made equal to 10,000 clock pulses and V_r equal to 1 V (+ or – as required), the count accumulated during discharge time (T_2) can be displayed digitally as the measured voltage.[17]

[17] "HPP3470 Digital Multimeter System," *Hewlett-Packard Journal*, Aug. 1972.

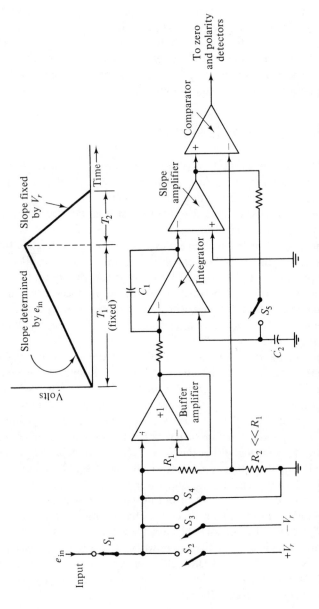

Figure 12.39 Dual-slope integration system used in a digital voltmeter (DVM); the input voltage e_{in} determines the height of the ascending ramp in the fixed time (T_1), while the descending ramp, initiated by reference voltage V_r, establishes the discharge time (T_2), which is proportional to the unknown e_{in}.

Advantages of Dual-Slope Method

This method, while retaining the basic advantage of an integrating-conversion method—in which *noise and other interference are averaged out*—also offers an important advantage in that other *potential error sources are self-canceling.* Where the accuracy of the single-integration method must suffer from any long-term changes in the clock rate or in the passive RC elements, such changes in the dual-slope method affect both up-and-down slopes equally, and therefore do not affect the count ratio.

In determining the overall accuracy of the digital voltmeter, it is necessary to consider other noncanceling errors, such as offsets in the ICs and nonlinear delays in the zero-detection process. Various methods are used to minimize such accuracy-degrading sources.[18] However, with the liberal use of the flexible and advanced ICs that are available, the net accuracy figures offered by even the relatively inexpensive DVMs are highly satisfactory, coupled with the very convenient small size that results from the high-count MSI and MOS/LSI forms of integration.

As a final example of the growing tendency for integrating more and more discrete functions on a single chip, a *single array that combines all the logic functions of a $4\frac{1}{2}$-digit voltmeter is the 3814 DVM array* illustrated in Fig. 12.40. This is an MOS type of digital IC (*Fairchild model A7R-3814-19X*).

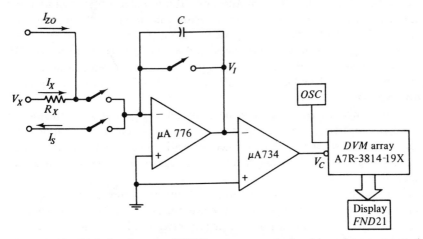

Figure 12.40 Digital voltmeter (DVM) array on a single chip; when connected to three OP AMPs, as shown in functional form, it operates as a complete DVM, with an LED (FND21) display unit. (*Fairchild Camera and Instrument Corporation*)

[18] Further details of DVM operation can be obtained from "HP3470 Digital Multimeter System," and also from S. Prensky, *Electronic Instrumentation*, 2nd ed., Prentice-Hall, Inc., Englewood Cliffs, N.J., 1971, Chap. 17.

The complete DVM circuit is formed when the array is combined with the three linear ICs shown (776 type OP AMP as integrator, 734 comparator type, and an *IC oscillator*). The output of the 3814 DVM array then feeds the multiplex display module (*FND21*) for the digital readout.[19]

In summary, in the *easy-to-read DVM*, we can also observe the great improvement in the *convenient compactness* and the *enhanced reliability and accuracy* that results from the strategic use of the combination of linear and digital ICs. While the pointer type (or analog type of meter), with its usual ±2 percent accuracy, still has its valid uses, the digital type of instrument—employing a number of ICs— has achieved a firm place for itself, spreading over the entire field of electronics.

12-15 SOURCES FOR SPECIFIC TECHNICAL INFORMATION

When one considers the rapid pace at which developments in linear ICs have been made and how many new models have been (and still are being) introduced as a result, it becomes clear that one of the most helpful aspects of a LIC manual will be to supply a convenient guide to reference material for further technical details of specific models. This has been done in this text in two ways: first, a cross-referenced index of LICs in Appendix III lists and identifies well some 400 model types in alphanumeric sequence, with a notation for the section where most of these can be found as discussed in the text. Second, at the end of this chapter we give a listing of references, and attention is called to the comprehensive commercial LIC manuals and application notes offered at no or slight cost by the semiconductor manufacturers (whose addresses are given in Appendix IV).

It is well to reemphasize a major objective of this text in treating the linear ICs separately from the even greater welter of digital ICs that are available. This follows the current practice of many semiconductor manufacturers, who now issue separate manuals for each category of discrete, digital IC, and linear IC devices.

A further subdivision, separating the OP AMPs from other LICs, is presented in the form of a *Quick Reference Selection Guide for Op Amps*, in Appendix II. Admittedly a partial listing, it serves as a further *simplification for initial identification* of the many available types that are typical of this versatile LIC device. In this connection, the *caveat* bears repeating that many additional devices are offered by *other manufacturers, whose names and addresses are listed* in Appendix IV. Much valuable technical data and applications may be gained from their informative catalogs.

[19] Fairchild 3814 data sheet and application note.

QUESTIONS

12-1. The designation of *linear IC* also includes such selected devices as
(a) Active filters
(b) Gates
(c) Flip flops

12-2. In programmable amplifiers (such as OTAs), the function of the R_{set} (or I_{set}) terminal varies the characteristics of the OP AMP by
(a) Varying the feedback ratio
(b) Varying the bias current (I_B)
(c) Varying the supply voltage (V_{CC})

12-3. In the operational transconductance amplifier (OTA) the change in g_m is produced
(a) By selection of feedback resistor
(b) By selection of load resistor
(c) By selection of R_{set}

12-4. Active filters are used (rather than the passive type) in order to
(a) Avoid the use of inductors in *RLC* circuits
(b) Avoid attenuation of signal in *RC* circuits
(c) Avoid multiple break frequencies

12-5. Multiple OP AMPs in a monolithic package are available only as
(a) Dual OP AMPs
(b) Triple OP AMPs
(c) Quad OP AMPs
(d) All of the above

12-6. In developing a function generator from two OP AMPs (as in Fig. 12.28(b)), the triangular wave output is developed from
(a) Integrating a square wave
(b) Clipping a sine wave
(c) Combining two ramp voltages

12-7. The advantage of the use of a dual-slope method for the DVM circuit (Fig. 12.39) lies in the fact that the resultant readout
(a) Is obtained in faster time
(b) Is independent of clock frequency for T_1 and T_2
(c) Is independent of random noise voltages

PROBLEMS

12-1. If, for the voltage regulator of Fig. 12.2, $V_{ref} = 10$ V and $R_1 = 10$ kΩ, what value is required of R_2 if the output voltage is 20 V?

12-2. Repeat Problem 12-1 for an output voltage of 28 V.

12-3. For the passive low-pass filter of Fig. 12.4(a), $RC = 1$ μs. Calculate the value of the cutoff frequency.

12-4. For the modified lag network of Fig. 12.4(b), $f_{c1} = 100$ Hz and $f_{c2} = 500$ Hz; $R_2 = 1$ kΩ. Calculate the values of R_1 and C.

12-5. Referring to Fig. 12.5(a), assume that $R_1 = R_2 = 30$ kΩ and the cutoff frequency is 20 kHz. Determine the values of C_1 and C_2.

12-6. Calculate the value of the RC product if the center frequency is 1 kHz for the band-pass active filter of Fig. 12.7.

12-7. It is desired to simulate an inductance of 2 H with the gyrator of Fig. 12.12. If all the resistors are each equal to 20 kΩ, determine the value of C_1.

12-8. For the state-variable filter of Fig. 12.9(a), $H_o = 10$ for the BP, HP, and LP outputs. Calculate the values of R_Q and R_{IN} for the following cases:
(a) BP output ($Q_0 = 10$)
(b) HP output
(c) LP output

12-9. Used as a gyrator, the GIC of Fig. 12.14 is connected to realize an inductance of 1.5 H. Calculate the value of C_2.

12-10. Referring to Fig. 12.15, $\omega = 1,000$ rad/s, $C_1 = 0.01$ μF, and $Z_i = -1,000$ Ω. Calculate the value of C_5.

12-11. For the values given in Fig. 12.28 for the Wien-bridge oscillator, calculate the frequency of oscillation.

12-12. If $R_1 = 10$ kΩ and $f_o = 2$ kHz, determine the value of C_1 for the VCO of Fig. 12.29.

12-13. The length of the output pulse required of the one-shot of Fig. 12.33 is 15 ms. If $R = 1.5$ kΩ, what value of C is required?

12-14. For the astable multivibrator of Fig. 12.34, $f = 1,000$ Hz. If $R_A = R_B$ and $C = 0.2$ μF, calculate the values of R_A and R_B.

12-15. Repeat Problem 12-14 for a frequency of 1,800 Hz.

12-16. If, in Fig. 12.36(a), $X = 3$ V and $Y = 5$ V, calculate the value of v_o.

12-17. Using the circuit of Fig. 12.36(b), what value of X is required to yield an output voltage of 1 V?

12-18. Referring to Fig. 12.37(a), if $Z = 1.3$ V and $X = 3$ V, what is the output voltage?

12-19. The output for the circuit of Fig. 12.37(b) is -3.4 V. What is the value of Z?

Bibliography

Barna, A., *Operational Amplifiers*, John Wiley & Sons, Inc. (Interscience Division), New York, 1971.

Coughlin, R. C., and F. F. Driscoll, *Operational Amplifiers and Linear Integrated Circuits*, Prentice-Hall, Inc., Englewood Cliffs, N.J., 1977.

Deboo, G. J., and C. N. Burrous, *Integrated Circuits and Semiconductor Devices*, 2nd edition, McGraw-Hill Book Company, New York, 1977.

Faulkenberry, L. M., *An Introduction to Operational Amplifiers*, John Wiley & Sons, Inc., New York, 1977.

Garrett, P. H., *Analog Systems for Microprocessors and Minicomputers*, Reston Publishing Company, Reston, Va., 1978.

Graeme, J. G., *Applications of Operational Amplifiers: Third Generation Techniques*, McGraw-Hill Book Company, New York, 1973.

Graeme, J. G., *Designing with Operational Amplifiers: Applications and Alternatives*, McGraw-Hill Book Company, New York, 1977.

Grinich, F. H., and H. G. Jackson, *Introduction to Integrated Circuits*, McGraw-Hill Book Company, New York, 1975.

Hnatek, E. R., *User's Handbook of Integrated Circuits*, John Wiley & Sons, Inc. (Interscience Division), New York, 1973.

Johnson, D. E., and J. L. Hilburn, *Rapid Practical Designs of Active Filters*, John Wiley & Sons, Inc. (Interscience Division), New York, 1975.

Kahn, M., *The Versatile Op Amp*, Holt, Rinehart and Winston, Inc., New York, 1970.

Kaufman, M., and A. H. Seidman, *Handbook of Electronics Calculations for Engineers and Technicians*, McGraw-Hill Book Company, New York, 1979.

Kaufman, M., and A. H. Seidman, *Handbook for Electronics Engineering Technicians*, McGraw-Hill Book Company, New York, 1976.

Lenk, J. D., *Handbook of Digital and Linear Integrated Circuits*, Reston Publishing Company, Reston, Va., 1973.

Lenk, J. D., *Handbook of Simplified Solid-State Circuit Design*, Prentice-Hall, Inc., Englewood Cliffs, N.J., 1971.

Melen, R., and H. Garland, *Understanding IC Op Amps*, Howard W. Sams Company, Inc., Indianapolis, 1971.

Millman, J., *Microelectronics*, McGraw-Hill Book Company, New York, 1979.

Moschytz, G. S., *Linear Integrated Networks Fundamentals*, Van Nostrand Reinhold Company, New York, 1974.

Roberge, J. K., *Operational Amplifiers: Theory and Practice*, John Wiley & Sons, Inc., New York, 1975.

Rutkowski, G. B., *Handbook of Integrated-Circuit Operational Amplifiers*, Prentice-Hall, Inc., Englewood Cliffs, N.J., 1975.

Seidman, A. H., and J. L. Waintraub, *Electronics: Devices, Discrete and Integrated Circuits*, Charles E. Merrill Publishing Company, Columbus, Ohio, 1977.

Smith, J. I., *Modern Operational Circuit Designs*, John Wiley & Sons, Inc., New York, 1971.

Stout, D. F., and M. Kaufman, *Handbook of Operational Amplifier Circuit Design*, McGraw-Hill Book Company, New York, 1976.

Tobey, G. E., L. P. Huelsman, and J. G. Graeme, *Operational Amplifiers, Design and Applications*, McGraw-Hill Book Company, New York, 1971.

Wait, J. V., L. P. Huelsman, and G. A. Korn, *Introduction to Operational Amplifier Theory and Applications*, McGraw-Hill Book Company, New York, 1975.

Wojslaw, C., *Integrated Circuits: Theory and Applications*, Reston Publishing Company, Reston, Va., 1978.

Commercial LIC Manuals and Application Notes
Virtually every manufacturer of LICs provides manuals at no, or nominal, cost and application notes. For inquiries, manufacturers' addresses are listed in Appendix IV.[1]

---

[1] A comprehensive listing of practically all available linear ICs is offered (on a two-issue-per-year subscription basis) as the *Linear IC D.A.T.A. Book*, from D.A.T.A. Inc., 32 Lincoln Ave., Orange, New Jersey 07050.

Appendix I

Decibel Conversion Chart

DECIBEL CONVERSION CHART FOR APPROXIMATE VOLTAGE-GAIN RATIO

Voltage-gain definition: No. of decibels (dB) $= 20 \log \dfrac{V_{out}}{V_{in}}$

No. dB		Numerical Gain Ratio	No. dB	Numerical Gain Ratio
Reference	0	= 1.0		
	1	= 1.12	13	= 4.47
	2	= 1.25	14	= 5.0
	3	= 1.41	15	= 5.62
	4	= 1.58	16	= 6.31
	5	= 1.78	17	= 7.08
	6	= 2.0	18	= 8.0
	7	= 2.24	19	= 8.91
	8	= 2.51	20	= 10.0
	9	= 2.82	40	$= 10^2 = 100$
	10	= 3.16	60	$= 10^3$
	11	= 3.55	80	$= 10^4$
	12	= 4.0	100	$= 10^5$

DECIBEL CONVERSION TABLE

Note: The decibel values were historically employed to express *power ratios;* but since then they have come into common use to express equivalent *voltage ratios, where, it should be noted, the formula relationship is different from the power ratio relationship.*

Since the decibel values in this text are used practically exclusively to express *voltage* amplification ratios, only these equivalent ratios[1] are given. The distinction is shown by the following definitions:

For voltage: no. dB = 20 log V_{out}/V_{in}.

For power: no. dB = 10 log P_{out}/P_{in}.

Therefore, it is important to consider *all the decibel references in the text as referring to voltage ratios, unless a power ratio is specifically mentioned.*

MENTAL CALCULATION OF GAIN

For any statement of voltage gain given in decibels, the numerical ratio can easily be estimated *mentally by remembering just two rules:*

Decibels		Numerical Gain Ratio
1. No. dB/20	=	power of 10
2. Adding dB as follows:	=	multiplying the ratio as follows:
+6 dB	=	2.0X
+3 dB	=	1.4X
+2 dB	=	1.25X
+1 dB	=	1.12X

EXAMPLE FOR 42 dB

1. Taking the nearest multiple of 20:
   40 dB/20 = 10^2

2. Adding 2 dB = 1.25X
   Then 42 dB = 100(1.25), or 125:1 as the gain ratio.

[1]Those equivalents that are printed in boldface are emphasized for convenience for mentally estimating the approximate numerical gain ratio that corresponds to a given decibel value.

Appendix II

Classification of LIC Groups and Selection Guide for Op Amps

CLASSIFICATION OF LIC GROUPS

Arrays (Chapter 2)

Audio power amps (Chapter 7)

Comparators (Chapter 10)

Consumer circuits (Chapter 8)

Digital interfacing (Chapter 10)

Instrumentation amplifiers (Section 11-9)

Multipliers (Section 12-13)

OP AMPs (See following selection guide)

Phase-locked loops (Section 8-8)

Regulators (Chapter 9)

Specialized systems (Chapter 12)

OP-AMP SELECTION GUIDE

1. General-Purpose Types[1]
   A. Internally compensated: LM107, MC1556, μA741
   B. Extended bandwidth: LM101/A, μA748, SN52660
   C. FET input: AD503, LH356, μA740
   D. Dual types: MC1558, μA747
   E. Quad types: CA3060 (triple), HA2400/2405, MC3301/P, LM3900

[1] Table 1-1 (Sec. 1-7) identifies "second-source" types; Table 4-1 (Sec. 4-6) gives comparative data on general-purpose OP-AMP types.

2. High-Performance Types[2]
    A. Wide band (fast slew rate): AD507, LM118, CA3100, NE/SE530, 1321(T/P)
    B. Low-bias (I_B), (FET input or superbeta): AD523, BB3521, LM108/A, MC1556, NE/SE536
    C. High-output current: AD517, μA791, CA3047, MC1538
    D. Low power (micropower) and operational transconductance amplifier (OTA): CA3080, μA776, LM112
    E. Precision (low drift and high accuracy): AD234, μA777, CA6078, SN52770/771
    F. Instrumental amplifiers: See Section 11-9.

---

[2] Table 11-1 (Sec. 11-3) gives comparative data on high-performance OP-AMP types.

Cross-Reference List of LICs and Key to Identification of Manufacturers' Types

CROSS-REFERENCE LIST OF LINEAR INTEGRATED CIRCUITS

Key to Identification: gives manufacturers' names (see addresses in Appendix IV).

Order of Listing: first, alphabetically by manufacturers' letter prefixes: e.g., MC1303, MC1304 . . . MFC . . . μA (μA = muA) . . . , followed by numerical designations.

Italic Numbers: represent much-used type numbers also appearing in "second-sourced" types; e.g., SN52*741* is equivalent to 741 type.

Section Numbers: refer to discussion in text.

KEY TO IDENTIFICATION OF MANUFACTURERS' TYPES

(See Appendix IV for Manufacturers' Addresses)

AD: Analog Devices
ADC-EK: Datel/Intersil
BB: Burr-Brown

CA ⎫
CD ⎬ RCA
CMP: Precision Monolithics

DAC: Hybrid Systems

DF: Siliconix

HA: Harris

HDAS: Datel/Intersil

HI: Harris

ICL ⎫
IH ⎬ Intersil

L: Siliconix

LAS: Lambda

LF ⎫
LH ⎬ National Semiconductor
LM ⎭

MC ⎫
MFC ⎬ Motorola
MPQ ⎭

μA: Fairchild

NE   ⎫
NE/SE ⎬ Signetics

RC ⎫
RM ⎬ Raytheon

S: American Microsystems

SG: Silicon General

SMP: Precision Monolithics

SN: Texas Instruments

TDA: Motorola

TL: Texas Instruments

ULN: Sprague

XR: Exar

ZN: Ferranti

276-series: Radio Shack

Note: Manufacturers' names are directly supplied for all numbered (no letter prefix) designation of devices.

Manufacturers' Type Number	Description	Section in Text
AD*101/A*	Op amp (extended bandwidth)	See LM*101/A*
AD234	Op amp (chopper stabilized)	11-5
AD503	Op amp (FET input, high accuracy)	11-4, 11-5
AD504	Op amp (precision, ultrastable)	
AD506	Op amp (FET input, low drift)	11-3
AD507	Op amp (high slew rate)	11-7
AD517	Op amp (high output current)	7-2, 11-3, 11-5
AD518	Op amp (wide band)	See LM*118/318*
AD520	Op amp (instrumentation)	11-10
AD523	Op amp (electrometer)	12-8
AD528	Op amp (FET input, high slew rate)	12-9
AD530	Multiplier	12-12
AD571	10-bit success. approx. ADC	2-17, 10-6
AD*741*	Op amp (internally compensated)	See μA*741*
AD2026	3-digit DPM	2-17, 5-18
ADC-EK12B	12-bit integrating ADC	
BB3329/03	Power Booster	7-3
BB3500	Op amp (low drift)	
BB3505/7	Op amp (high slew rate)	
BB3506/8	Op amp (wide band)	
BB3521	Op amp (electrometer)	11-5, 12-8
BB3522	Op amp (low offset voltage)	
BB3524	Op amp (FET input)	
BB3542	Op amp (FET input)	
BB3554	Op amp (high slew rate)	11-7

Manufacturers' Type Number	*Description*	*Section in Text*
BB3625/A	Instrumentation amp (variable-gain module)	
BB3630	Instrumentation amp (thin-film)	11-11
BB4127	Log amp (hybrid)	12-7
CA2111/A	FM-IF amplifier; limiter/detector	
CA3005	RF/IF amplifier (similar to μA703)	
CA3008/16	Op amp (wide band)	
CA3010/15	Op amp (wide band)	
CA3014	TV sound IF amp and detector	
CA3018/A	Quad NPN transistor array	2-11
CA3019	Quad diode array	
CA3020/A	Audio/video power amplifier (1 W)	
CA3026	Dual RF/IF differential amplifier array	2-11
CA3028/A/B	Differential cascode amp (to 120 Hz)	
CA3029/30	Op amp (wide band)	
CA3030/A	Op amp (low noise wide band)	
CA3033/A	Op amp (high output current)	
CA3035	Triple wide band amp	
CA3036	Dual Darlington array	
CA3037/38	Op amp (wide band)	
CA3039	6-diode array	
CA3043	FM-IF amplifier/limiter	8-3
CA3045/6	5-transistor array	2-11
CA3047/A	Op amp (high output current)	
CA3048	Quad amplifier array	2-15
CA3049	Dual differential amplifier array	
CA3050	Dual Darlington diff amp array	
CA3053	Diff cascode amp (to 10.7 MHz)	
CA3054	Dual differential amplifier array	
CA3059	Zero-voltage switch	9-8
CA3060	Triple OTA	12-10
CA3062	Photodetector and power amplifier	
CA3064	TV-AFT control	8-6
CA3065	TV-FM sound system	8-6
CA3066	TV chroma amplifier	8-6
CA3067	TV chroma demodulator	8-6
CA3068	TV (video IF and AM-IF)	8-6
CA3070	Chroma signal processor (PLL)	
CA3071	Chroma amplifier	
CA3072	Chroma demodulator (see also μA746)	
CA3075	FM radio system	
CA3078	OTA (programmable)	11-8, 12-2
CA3080/A	OTA (programmable)	4-6, 11-8, 12-12, 12-13
CA3081	7-transistor common-emitter array	2-11
CA3082	7-transistor common-collector array	2-11
CA3083	5-transistor high current array	2-11
CA3084	General purpose PNP transistor array	2-11
CA3085	Positive voltage regulator (similar to LM 105)	

Manufacturers' Type Number	Description	Section in Text
CA3088	AM radio subsystem	
CA3089	FM radio subsystem	
CA3091	Analog multiplier	
CA3093	Transistor/zener diode array	2-11
CA3094/A	High-current output power switch	7-2, 9-7, 12-2
CA3095/A	Transistor array (super beta)	
CA3096/A	NPN/PNP transistor array	6-9
CA3097E	Thyristor/transistor array	2-11, 9-8
CA3100	Op amp (wide band)	6-9
CA3102	Dual high-frequency diff amp	
CA3118/A	4-transistor high-voltage array	2-11
CA3120	TV signal processor	
CA3123/E	AM radio subsystem	
CA3130/3140	Op amp (BiMOS)	5-17, 6-9
CA3139	TV-AFT control	8-6
CA3146/A	5-transistor high voltage array	2-11
CA3183/A	5-transistor high current array	2-11
CA3189E	IF, detector, pre-amp	8-3
CA3724/5	4-transistor high current array	2-11
CA6078	OTA (low-noise version of CA3078)	
CA6741	Op amp (low-noise version of 741)	12-8
CD4007/A	COS/MOS for DAC	10-7
CD4046A	PLL (micropower	
CMP-02	Voltage comparator (precision)	10-4
DAC372-12	Hybrid DAC	10-6
DF341	CODEC	
HA909/911	Op amp (low noise)	12-8
HA2050	Op amp (high slew rate)	
HA2060	Op amp (FET input, wide band)	
HA2400/2405	Quad op amp (gate controlled)	12-10
HA2500	Op amp (high slew rate, internally compensated)	
HA2530	Op amp (high slew rate)	6-9
HA2820	PLL (to 25 MHz)	12-11
HA2825	PLL (from 0.1 Hz to 3 MHz)	12-11
HA2900	Op amp (chopper stabilized)	11-5
HDAS16	16-bit data acquisition system	
HI5612	12-bit DAC	
ICL741HS	Op amp (741, higher slew rate)	
ICL8001	Comparator (precision)	10-4
ICL8007	Op amp (FET input, precision)	12-7
ICL8021	Op amp (programmable)	
ICL8038/A	VCO	12-11
ICL8048	Log amplifier (monolithic)	12-7
ICL8101	Op amp (extended bandwidth)	See LM101/A
ICL8741	Op amp (internally compensated)	See μA741
IH5009	Analog gate switch (15 V)	
IH5010	Analog gate switch (5 V)	
L144	Op amp (programmable)	

Manufacturers' Type Number	Description	Section in Text
LAS1500	Voltage regulator (positive)	9-7
LAS1800	Voltage regulator (negative)	9-7
LF353	Op amp (dual JFET input, wide band)	
LF356	Op amp (BiFET)	5-17
LH0001/A	Op amp (micropower)	11-7
LH0023	Sample and hold	12-9
LH0043	Sample and hold	12-9
LH0052	Op amp (electrometer)	12-8
LH101	Op amp (internally compensated)	12-4
LH*740*/A	Op amp (FET input)	See μA*740*
LM100	Voltage regulator (positive)	
LM*101*/A	Op amp (extended bandwidth)	4-6, 5-6, 5-8, 5-10, 6-5, 6-8
LM*102*	Op amp (voltage follower)	12-9
LM104	Voltage regulator (negative)	
LM105	Voltage regulator (positive)	
LM*106*	Voltage comparator (high speed)	10-4
LM*107*	Op amp (internally compensated)	12-9, 12-11, 12-12
LM*108*/A	Op amp (super beta)	1-7, 4-6, 5-7, 5-15, 6-4, 11-2, 11-4
LM*109*	Voltage regulator (5 V)	9-7
LM*110*	Voltage follower (high slew rate)	6-9, 12-5
LM111	Voltage comparator	10-4
LM112	Op amp (micropower)	
LM117/217/317	Voltage regulator (3 terminals, adjustable)	9-7
LM*118*	Op amp (high slew rate)	11-7
LM119	Voltage comparator (dual, high speed)	
LM121/A	Op amp (low drift)	
LM146/246/346	Quad op amp (programmable)	12-5
LM170	AGC/squelch amplifier	
LM171	RF/IF amplifier	
LM172/372	AM-IF strip	8-5
LM311	Voltage comparator (precision)	
LM317	Voltage regulator (adjustable)	
LM324	Quad op amp (single supply)	5-16
LM334	Current source (3-terminal, adjustable)	
LM339	Quad voltage comparator	10-4
LM373	AM/FM SSB-IF strip	
LM376	Voltage regulator (positive)	9-7
LM377	Dual power amp (2 W per channel)	7-5
LM378	Dual power amp (4 W per channel)	7-6
LM380	Power amp (2.5 W)	7-5
LM381/382	Dual stereo preamp	7-5
LM383	Power amp (8 W)	
LM386	Power amp (low voltage)	
LM*565*	PLL	See NE/SE*565*
LM567	Tone decoder	

Manufacturers' Type Number	Description	Section in Text
LM709	Op amp	See µA709
LM710	Voltage comparator	See µA710
LM711	Voltage comparator (dual)	See µA711
LM723	Voltage regulator	See µA723
LM741	Op amp (internally compensated)	See µA741
LM747	Op amp (dual 741 type)	See µA747
LM748	Op amp (extended bandwidth,	See µA748
LM1303	Dual stereo preamp	See MC1303
LM1558	Dual op amp	See MC1558
LM1877	Dual 2-W power amplifier	
LM2111	FM-IF amplifier (limiter/detector)	See CA2111/A
LM3028/A/B	Differential cascode amp	See CA3028/A/B
LM3064	TV-AFT control	See CA3064
LM3900	Quad op amp	12-10
LM4250	Op amp (programmable)	
LM7524	Dual sense amplifier	See SN7524
LM7525	Dual sense amplifier	See SN7525
MC1303	Dual stereo preamp	7-7
MC1305	Stereo FM multiplex decoder	8-4
MC1306	Power amplifier (0.5 W)	7-5
MC1310	FM stereo demodulator	8-4
MC1312	Quadraphonic decoder (SQ matrix)	
MC1324	TV chroma demodulator	
MC1351	TV sound system	
MC1357	IF amp and quadrature detector	See CA2111/A
MC1358	TV (sound IF)	See CA3065
MC1364	FM-AFT control	
MC1391P	Positive flyback, hor sweep	8-6
MC1393	Vertical sweep	8-6
MC1394P	Negative flyback, hor sweep	8-6
MC1398	TV chroma signal processor	
MC1438R	Power booster	7-3
MC1445/1545	Op amp (differential output)	12-10
MC1458	Dual op amp	
MC1488	Quad line driver	
MC1489	Quad line receiver	
MC1508	Multiplying DAC	10-6
MC1514	Dual differential comparator	10-4
MC1530	Op amp (general purpose)	
MC1536	Op amp (high voltage)	
MC1537	Op amp (dual MC1709)	
MC1538	Power booster	7-3
MC1539	Op amp (fast slew rate)	
MC1540	Core memory sense amplifier	
MC1545	Op amp (wideband, gate control)	12-10
MC1550	RF/IF amplifier	8-5
MC1553	Video amplifier	
MC1554	Power amplifier (1 W)	7-5

Manufacturers' Type Number	Description	Section in Text
MC*1556*	Op amp (super beta)	11-2, 11-4
MC*1558*	Op amp (dual MC1741)	See μA*747*
MC1563	Voltage regulator (negative)	
MC1566	Regulator (voltage and current)	9-7
MC1568	Voltage regulator (dual polarity)	9-7
MC1569	Voltage regulator (adjustable)	
MC*1595*	4-quadrant analog multiplier	8-7
MC*1596*	Double-balanced mixer/modulator	8-7, 12-13
MC1*709*	Op amp (general purpose)	See μA*709*
MC1*710*	Differential voltage comparator	See μA*710*
MC1*711*	Comparator (dual)	See μA*711*
MC1*723*	Voltage regulator (adjustable)	See μA*723*
MC1*733*	Differential video amplifier	See μA*733*
MC1*741*	Op amp (internally compensated)	See μA*741*
MC1*748*	Op amp (extended bandwidth)	See μA*748*
MC3301/P	Quad op amp (single supply)	12-10
MC3302/P	Quad comparator	10-4
MC3357	Narrow-band FM-IF strip	8-3
MC3430	Quad comparator	10-4
MC3476	OTA (programmable)	11-8
MC4024/4324	VCO	12-11
MC4044/4344	Phase/frequency detector	12-11
MC4344	PLL	12-11
MC14443	Ramp-type ADC	10-6
MC14447	Ramp-type ADC	10-6
MFC4050	Power amplifier (4 W)	
MFC6010	FM-IF amplifier	
MFC8010	Audo amp (1 W, including preamp)	7-5
MFC8020/A	Audio class B driver	
MFC8040	Preamplifier (low noise)	
MFC9020	Power amplifier (2 W)	7-5
MPQ6502	NPN/PNP high-current array	2-15
μA78S40	Switching voltage regulator	9-7
μA*702*	Op amp (wide band)	
μA*703*	FM-RF/IF amplifier	8-2
μA*706*	Power amplifier (5 W)	7-6
μA*709*	Op amp (general purpose)	4-6, 11-7
μA*710*	Voltage comparator	10-3
μA*711*	Voltage comparator (dual)	10-4, 10-5
μA715	Op amp (high slew rate)	11-7
μA720	AM subsystem	
μA*723*	Voltage regulator	9-5
μA*725*	Op amp (precision)	6-7
μA727	Temperature-controlled preamp	
μA729	FM stereo MPX decoder	
μA*730*	Differential amplifier	
μA732	FM stereo MPX decoder	
μA*733*	Differential video amplifier	
μA734	Voltage comparator (precision)	10-4, 10-6

Manufacturers' Type Number	Description	Section in Text
μA735	Op amp (micropower)	
μA739	Op amp (dual, low noise)	6-9
μA740	Op amp (FET input)	4-6, 6-9, 11-2, 11-4
μA741	Op amp (internally compensated)	1-7, 4-2, 4-4, 4-6, 5-4, 5-9, 5-11, 5-13, 5-14, 6-3, 6-4, 6-8
μA742	Zero-crossing ac trigger	
μA745	Dual ac amplifier	
μA746	TV chroma demodulator	See CA3072
μA747	Op amp (dual μA741)	4-6, 12-6
μA748	Op amp (extended bandwidth)	4-5, 4-6, 5-5
μA749	Dual preamplifier	
μA750	Voltage comparator (dual)	10-4
μA753	FM gain block	
μA754	TV/FM sound system	
μA757	AGC-IF amplifier	
μA758	Stereo decoder (PLL)	8-4
μA760	Voltage comparator (high speed)	4-6, 5-8, 6-8, 10-4
μA761	2-channel sense amplifier	
μA767	Stereo decoder	
μA768	Stereo decoder	
μA769	Stereo decoder	
μA770	Op amp (super beta)	4-6
μA771	Instrumentation amplifier	11-10
μA772	Op amp (high slew rate)	11-7
μA776	Op amp (programmable)	11-8, 12-2
μA777	Op amp (precision)	4-6, 5-8, 5-12, 11-3
μA780	Chroma signal processor (PLL)	See CA3070
μA781	TV chroma amplifier	See CA3071
μA791	Op amp (high output current)	7-2
μA795	4-quadrant multiplier	
μA796	Double-balanced modulator/demodulator	
μA78XX	Voltage regulator (positive; 7805 = 5 V up to 7828 = 28 V; preset output)	
μA9709	12-bit DAC	
NE511	Dual differential amplifier	3-5
NE546	AM receiver subsystem	8-5
NE/SE155	Op amp (BiFET)	5-17
NE/SE259	Voltage comparator (precision)	10-4
NE/SE515	Op amp (diff input and output)	
NE/SE527	Voltage comparator	
NE/SE530	Op amp (high slew rate)	
NE/SE536	Op amp (FET input)	
NE/SE540	Power driver (1 W)	
NE/SE550	Voltage regulator (precision)	

Manufacturers' Type Number	Description	Section in Text
NE/SE555	Timer	12-12
NE/SE556A	Dual 555 timer	12-12
NE/SE562	PLL	12-11
NE/SE565	PLL	8-8
NE/SE566	VCO	12-11
NE/SE567	Tone decoder (PLL)	8-8
NE/SE5540	Instrumentation amplifier	2-17
NE/SE52660	Op amp (precision)	4-6
RC105/A	Voltage regulator (precision positive)	See LM105/A
RC108/A	Op amp (super beta)	See LM108/A
RC111	Voltage comparator	See LM111
RC118	Op amp (high slew rate)	See LM118
RC723	Voltage regulator	See μA723
RC748	Op amp (extended bandwidth)	See μA748
RC1556/A	Op amp (super beta)	See MC1556
RC4136	Quad (741) op amp	See LM3900
RM101/A	Op amp (extended bandwidth)	See LM101/A
RM4131	Op amp (precision)	
RM4132	Op amp (micropower)	
RM4531	Op amp (high slew rate)	
S2814	Fast Fourier transformer	
SG1524/3524	Regulating pulse width modulator	
SG3523/3423	Voltage sensor	
SMP11	Sample and hold amp	12-9
SN62/72088	Op amp (chopper stabilized)	11-5
SN7524	Dual sense amplifier	10-5
SN7525	Dual sense amplifier	10-5
SN52660	Op amp (precision)	4-6, 6-8
SN52741	Op amp (internally compensated)	1-7
SN52770	Op amp (super beta)	4-6, 6-8
SN52771	Op amp (super beta)	11-3
SN76011	Power amplifier (1 W)	7-5
SN76024	Power amplifier (4 W)	7-6
SN76242	Chroma signal processor (PLL)	See CA3070
SN7623	Chroma amplifier	See CA3071
SN76267	TV chroma demodulator	See CA3067
SN76477	Complex sound generator	
SN76564	TV-AFT control	See CA3064
SN76665	TV-FM sound system	See CA3065
TDA2002	Power amp (8 W)	7-6
TL061/071/081	Op amp (BiFET)	5-17
TL084CN	Op amp (JFET Input)	
TL480	Log 10-step analog level detector	
TL487	Log 5-step analog level detector	
TL489	Linear 5-step analog level detector	10-5
TL490	Linear 10-step analog level detector	
ULN2111/A	Quadrature FM detector	8-7
ULN2280B	Power amplifier (2 W)	7-5
XR084	Quad BiFET op amp	
XR205	Waveform generator	12-11

Manufacturers' Type Number	Description	Section in Text
XR215	PLL	
XR2207	VCO	12-11
XR2208/2308	Operational multiplier	12-12
XR2211	FSK demodulator (PLL)	8-8
ZN414	10-transistor TRF circuit	8-1
276-007	Op amp (internally compensated)	See μ741
276-038	Dual op amp	See MC1458
276-702	Dual 2-W power amplifier	See LM1877
276-703	Power amplifier (8 W)	See LM383
276-1711	Quad op amp	See LM324
276-1712	Quad voltage comparator	See LM339
276-1713	Quad OTA	See LM3900
276-1714	Op amp (JFET input)	See TL084CN
276-1715	Op amp (dual, JFET input)	See LF353
276-1721	Tone decoder	See LM567
276-1723	Timer	See NE/SE555
276-1724	VCO	See NE/SE566
276-1728	Dual 555 timer	See NE/SE556
276-1731	Power amplifier (low voltage)	See LM386
276-1734	3-terminal adjustable current source	See LM334
276-1740	Voltage regulator	See μA723
276-1765	Complex sound generator	See SN76477
276-1770	Voltage regulator (5 V)	See μA7805
276-1771	Voltage regulator (12 V)	See μA7812
276-1772	Voltage regulator (15 V)	See μA7815
276-1777	3-terminal adjustable voltage regulator	See LM317
276-1790	Volt-to-freq/freq-to-volt converter	See 9400
276-2446	PLL (micropower)	See CD4046A
840	Voltage reference, precision (Beckman/Helipot)	9-7
881	Universal active filter, hybrid (Beckman/Helipot)	12-6
1319	Op amp, low drift (Teledyne/Philbrick)	
1321	Op amp, wide band (Teledyne/Philbrick)	
1322	Op amp, high slew rate (Teledyne/Philbrick)	
1323	Op amp, micropower (Teledyne/Philbrick)	
4253	Instrumentation amp, modular FET input (Teledyne/Philbrick)	11-11
4450	4-quadrant multiplier, module (Teledyne/Philbrick)	12-13
4452	Analog multiplier, module (Teledyne/Philbrick)	12-13
7541	Multiplying 12-bit DAC (Beckman/Helipot)	
9400	Volt-to-freq/freq-to-volt converter (Teledyne Semiconductor)	

Appendix IV

Manufacturers' Addresses

American Microsystems, Inc., 3800 Homestead Road, Santa Clara, CA 95051.

Analog Devices, P.O. Box 280, Route 1 Industrial Park, Norwood, MA 02062.

Beckman Instruments, Inc., Helipot Division, 2500 Harbor Blvd., Fullerton, CA 92634.

Burr-Brown Research Corp., International Airport Industrial Park, Tucson, AZ 85734.

Datel-Intersil, 11 Cabot Blvd., Mansfield, MA 02048.

Exar Integrated Systems, 750 Palomar Ave., Sunnyvale, CA 94086.

Fairchild Camera and Instrument Corp., 464 Ellis St., Mountain View, CA 94042.

Ferranti Electric, Inc., 87 Modula Ave., Commack, NY 11725.

Harris Semiconductor, P.O. Box 883, Melbourne, FL 32901.

Hybrid Systems, Crosby Drive, Bedford Research Park, Bedford, MA 01730.

Intersil Inc., 10900 N. Tantau Ave., Cupertino, CA 95014.

Lambda Semiconductor, 121 International Drive, Corpus Christi, TX 78410.

Motorola Semiconductor Products, 5005 E. McDowell Road, Phoenix, AZ 85008.

National Semiconductor, 2900 Semiconductor Drive, Santa Clara, CA 95051.

Precision Monolithics Inc., 1500 Space Park Drive, Santa Clara, CA 95050.

Radio Shack, A Division of Tandy Corp., 1400 One Tandy Center, Fort Worth, TX 76102.

Raytheon Co., Semiconductor Division, 350 Ellis St., Mountain View, CA 94042.

RCA Solid State Division, Box 3200, Sommerville, NJ 08876.

Signetics Corp., 811 E. Arques Ave., P.O. Box 409, Sunnyvale, CA 94086.

Silicon General, Inc., 11651 Monarch St., Garden Grove, CA 92641.

Siliconix Systems, Inc., 14351 Myford Road, Tustin, CA 92680.

Sprague Electric Co., Semiconductor Division, 115 Northeast Cutoff, Worcester, MA 01606.

Teledyne Philbrick, Allied Drive at Route 128, Dedham, MA 02026.

Teledyne Semiconductor, 1300 Terra Bella Ave., Mountain View, CA 94043.

Texas Instruments, Inc., Semiconductor Group, P.O. Box 225012, MS 308, Dallas, TX 75265.

Appendix V

Answers to Selected Questions and Problems

ANSWERS TO QUESTIONS

2-1. (a), (d), (e), (f)
2-2. (c)
2-5. (a), (b), (d)
2-6. (c)
2-7. (a)
2-10. (a), (b), (c)

3-1. (c)
3-2. (c)
3-4. (b)
3-6. (b)
3-7. (d)
3-9. (b)

4-1. (a)
4-3. (b)
4-5. (b)
4-7. (c)

ANSWERS TO PROBLEMS

3-1. (a) – 60
3-3. 2000, 66 dB
3-5. (a) 46 dB, 52 dB, 66 dB
3-7. (a) – 450, (b) – 900, (c) – 450

4-1. (a) 5 mV, (b) 3 mV, (c) 1 mV
4-2. (a) 220 kΩ, (b) 2.7 MΩ,
     (c) 47 kΩ, (d) 100 kΩ
4-3. (a) – 5000, (b) – 500

345

ANSWERS TO QUESTIONS

5-1.	(c)
5-3.	(b)
5-5.	(b)
5-6.	(a)
5-7.	(c)
5-9.	(a)

6-1.	(a)
6-2.	(a)
6-3.	(b)
6-5.	(a)
6-7.	(a)

7-1.	(a)
7-2.	(c)
7-4.	(b)
7-5.	(a)
7-7.	(a)

8-1.	(c)
8-2.	(b)
8-3.	(a)

9-1.	(c)
9-3.	(a)
9-4.	(c)
9-5.	(c)
9-7.	(a)
9-8.	(a)
9-9.	(c)

10-1.	(a)
10-2.	(b)
10-3.	(a)
10-4.	(c)
10-5.	(c)

11-1.	(c)
11-2.	(b)
11-4.	(c)

ANSWERS TO PROBLEMS

5-1.	-1.118 V
5-2.	1.148 V
5-3.	(a) -6 V, (b) -4 V
5-7.	0.5 s

6-1.	(a) 600 Ω, (b) 2 V
6-3.	(a) 526 Ω
6-5.	(a) 500 kΩ, (b) 300 kΩ
6-7.	144.54 Hz

7-1.	8 Ω
7-3.	(a) 2 W, (b) 11.3 V, (c) 16
7-4.	135 mW

8-1.	0.1, 2, 5, 10
8-2.	300 s
8-4.	$\pm 0.67 \times 10^{-3}$ Hz
8-5.	± 335 kHz

9-1.	1%
9-3.	0.33%
9-4.	11.88 V
9-5.	2.13 kΩ
9-7.	12.5 Ω

10-1.	11 V
10-3.	8.125 V

11-1.	350 kHz
11-2.	-10 V
11-4.	14

ANSWERS TO QUESTIONS   ANSWERS TO PROBLEMS

11-5. (a)	**11-7.** ∞, 20 Ω
11-7. (c)	

12-2. (b)	**12-1.** 10 kΩ
12-4. (a)	**12-3.** 159 kHz
12-5. (d)	**12-5.** $C_1 = 375$ pF, $C_2 = 188$ pF
12-6. (a)	**12-7.** 0.005 μF
12-7. (c)	**12-9.** 0.027 μF
	12-11. 79.5 kHz
	12-12. 0.0125 μF
	12-14. 2400 Ω
	12-17. 3.16 V
	12-19. 1.156 V

Index